Contents

1, Glacier National Park, K-0028, by Nancy Livingston, N9NCY. 2, Lone Tree Wildlife Management Area, K-7702, by Andrew "Jim" Danielson, AC9EZ. 3, Birch Point Beach State Park, K-2382, by Kevin Thomas, W1DED. 4, Monument-Lefebvre National Historic Site, VE-4814, by Pete Kobak, KØBAK. 5, Honeymoon Island State Park, K-1880, by Lisa Neuscheler, KC1YL. 6, Zebulon B. Vance Birthplace State Historic Site, K-6856, by Thomas Witherspoon, K4SWL.

The Parks on the Air Book

225 Main Street, Newington, CT 06111-1400 USA
www.arrl.org

Printed in the USA

ISBN: 978-1-62595-174-8

First Edition

About the Cover Photos

ARRL member Michael Martens, KB9VBR, of Wausau, Wisconsin, produces the YouTube channel "KB9VBR Antennas" (**youtube.com/@KB9VBRAntennas**). The photograph was taken at the Dells of Eau Claire county park in Marathon County, WI. The Ice Age National Scenic Trail, K-4238, runs through the park.

ARRL member Jherica Goodgame, KI5HTA, of Avon, Connecticut has served as a summer intern at ARRL Headquarters. She attends college at the University of Mississippi and participates in POTA. The photograph was taken at the John W. Kyle State Park, K-2540, in Sardis, MS.

3

Foreword

Imagine that it is 2359 Zulu on December 31, 2016. A new year is about to begin, but there are those of us for whom thoughts of 2017 are far away. Some of us are concentrating on logging that final contact before the clock strikes 12 and it all ends. Time marches on and the inevitable happens. The end of the 2016 ARRL National Parks on the Air (NPOTA) event has just occurred. Thousands of amateur radio operators around the globe feel stunned, saddened — perhaps they regret not getting involved earlier in the year.

By the numbers, this event, planned and executed by Sean Kutzko, KX9X, and Norm Fusaro, W3IZ, of ARRL, was monstrously successful. It brought hundreds if not thousands of dormant operators out of hibernation (including Rick Parent, WØZAP, one of the authors of this foreword) and was instrumental in introducing our hobby to the next generation. Never had we had this much fun with amateur radio. Never had we learned more about one of our country's fabulous resources: the National Park Service. Never had we conducted as many serious portable HF operations as we did that year.

The end of the event left many of us with a large hole in our hearts as we sat around asking the question, "What now?"

Persistence pays. A couple of months into 2017, a few of us, led by Jason Johnston, W3AAX, began formulating what we could then only imagine: a fully functional US program for portable amateur radio operations. Thus, the Parks on the Air program (POTA) was born.

The resource pool of talent in amateur radio is endless. POTA was able to draw on that pool and enlist the help of many to mold our program into what it is today. Programmers, database managers, server administrators, media gurus, graphic artists, and many others made huge contributions to the program. All voluntary. All for the simple love of the hobby.

Fast forward from that cold December evening in 2016. Turn on any amateur radio and scan the bands. Day or night, around the clock and around the globe, you'll likely hear a POTA activation. From meager beginnings, the POTA phenomenon has gripped thousands.

From Anchorage, Alaska to Cape Town,

South Africa; from Miami, Florida to Sydney, Australia, and all points in between. With over 60 active DX entities currently participating in the program, POTA has truly become a global activity.

As of this writing, there are over 36,000 registered users of POTA. There are over 420,000 unique call signs in the POTA database. We have logged well over 15 million contacts from over 11,000 activators and over 420,000 hunters. That number increases daily. Regardless of age or gender, and whether activating alone or as a youth group, a club, or on a family camping outing in the great outdoors — there are opportunities for everyone to participate in the program.

This book has been compiled from the experiences of those operators. It details the process of activating a park unit as well as the "chasing" or "hunting" of those activations. Here you will read stories of what makes a successful POTA activation as well as some pitfalls to avoid. We have all had those days where the best laid plans don't quite work the way we envision. You will be able to read opinions on what works and what does not. Much of ham radio is built around what people can design, devise, and make up on their own with limited resources. POTA is no exception. Satisfying participation in the program requires antenna construction that works, essential equipment, as well as operating procedures for success.

Interested in trying something different in the world of ham radio? Come join us and explore the possibilities of portable operation. POTA on!

Jason Johnston, W3AAX
Rick Parent, WØZAP
May 2023

PART 1

Welcome to the Park

Kinzua Bridge State Park, K-1366 – Warren County, Pennsylvania

Photo by Kody L. Beer, K3KLB

1 Welcome to POTA

Lessons from a Young Activator

By **Jherica Goodgame, KI5HTA**

My dad introduced me to amateur radio at a very young age, and because of that, I can hardly remember it being new. It has been a constant in my life. It never gave me that nervous-yet-exciting feeling that comes along with trying something for the first time. For me, that feeling came with Parks on the Air (POTA). I started activating a few years ago and, despite my nervousness, had an absolute blast. I was lucky to have my father as a mentor (see **Figure 1**). He helped me through many of my first *activations*. (*Editor's note*: An activation is when you go out to a park to make contacts. If you get 10 you have *activated* the park and you get credit for it on the POTA website.) He taught me to set up a portable station, operate, and upload POTA contacts. So, in the spirit of helping others try new things, I want to talk about some of the challenges and triumphs I've experienced, and what I've learned from them during POTA activations. I'll cover everything from getting out to a park to submitting the contacts back at home.

Before Heading Out

Registering an account on the POTA website is an important part of setup, one you should do before you even go out to the park. (*Editor's note:* For a quick rundown of registering on the POTA site, see Harold Kramer, WJ1B's, essay "How to Become a POTA Hunter" later in this book.) It's easy, and the website pretty much gives you all the instructions. You need an account for spotting, which will I'll explain shortly. Making contacts can also be pretty easy. If you like, you call CQ and round up your contacts from people scrolling through the frequencies just like you

Figure 1, Jherica Goodgame, KI5HTA, and her father, Steve Goodgame, K5ATA, activating a park.

would during normal HF operation. Or, if you want to increase your chances of getting the 10 contacts needed to activate a park, you can sign into your account on the POTA website.

Once you are all set up and out in the park, you will want to create a *spot* on the website **pota.app**. A spot gives hunters and other activators a way to see your frequency, what park you are at, your call, and they can even re-spot you. Re-spotting brings your spot back to the top of the page, so it reaches more people. Some of the coolest contacts to make are contacts with people doing other POTA activations, which are called *park-to-park* contacts. Setting up an account, in my opinion, is a must for activating — you reach so many more people that way.

I'd also suggest gathering all your gear in one place, even in a single bag. Having your gear spread out and unorganized can make for a more confusing experience when you arrive. I have a backpack with a lot of compartments to store all the stuff I use for POTA in one place, and it helps significantly. I never have to keep running back to the car or anything. I just take the backpack with me, and it is all right there to begin setup.

KI5HTA's Gear

- Icom IC-705 transceiver
- LiFePO4 battery (Bioenno)
- End-fed half-wave antenna
- iPad with *HAMRS* for logging
- Chips and drinks for snacking!

Activating a Park

When it comes to activating a park, getting the hang of the entire setup process is probably the most important part. Setting up your portable station can seem like a challenge at first, but once you get the hang of it, it can take about five minutes or less. Setup for me is relatively simple. I start by setting up whichever antenna I use, which is generally an end-fed half-wave (EFHW), so this step is fairly simple. I use nearby trees, but you can use a mast or a long fishing pole. Then I plug it into the radio, hook up the mic, and make sure everything sounds good. Then, I find an empty frequency and ask if it is in use. If I hear nothing, I start calling CQ! There is nothing extra you must do after setting up — throw up the antenna, plug everything in, and you're good to go! That's what makes portable operations so much fun. It's quick, easy, and — with POTA — a good way to connect with the environment while you operate.

However, there can be some challenges to a POTA activation. For example, low power can be a concern. I remember once when I was operating QRP using an Icom IC-705, and it was near impossible for anyone to hear me. It was frustrating. However, people who want to make a contact will try to hear you, and if they just can't, there is no shame in signing off and trying again later. After all, the point of POTA is to have fun. But of course, the radio is not the only factor at play in ham radio — you have things like the weather, the antenna, if there is a contest that day, and a variety of other things that can influence the entire experience. My advice for times when things don't work out? Stay positive. Remember that you are there to have fun, and don't be too hard on yourself. POTA is a fun and enriching pastime, but you cannot let something out of your control ruin your day. Remember, you can always try again later.

The memories you make from POTA activations definitely outweigh the activity's challenges. When I first started out (and still to this day when I do an activation), I pretty much always operated with my dad. My whole family — my dad, my mom, and I — would go out to a park, maybe take a walk to get to know it. We'd find a nice, flat place to set up (usually with an awesome view), and we'd operate. My dad and I would take turns, and we always ran a little contest to see who could make more contacts (spoiler: I always won — see **Figure 2**).

Memories like these are a constant in my activations, and they're my favorite part of POTA. Getting to connect with my family, with nature, and most of all, with so many different people out there all at the same time is quite possibly the coolest experience ever.

Figure 2, Operating voice while K5ATA logs contacts.

A Favorite POTA Memory

I remember every detail of one activation I had. I was with my family at a park we liked to activate, John W. Kyle State Park, K-2540, in Mississippi. Sun peeked through arrays of trees. Walking trails wound beside Sardis Lake. The whole park smelled of clean, clear water. It was perfect for POTA. We usually stationed ourselves at a picnic table with a view of the lake, so we could enjoy the breeze while we operated. On a sunny day with a portable setup, some snacks, and a cheery attitude, it was one of my favorite places to be. That day, when it was my turn on the radio, I had one of the biggest pileups I can ever remember working. I mean, I was waiting a solid 8 – 10 seconds for everyone to finish calling. There were a *lot* of people. About midway through, one elderly gentleman called me, and I heard him. So, I made a contact. Usually, in a pileup contacts would go pretty quickly. I would get the contact's call sign, give and receive a signal report, tell them to have a nice day, 73, QRZ, and on to the next. It was a pretty quick process. Generally, people could hear the pileup, so they understood. This gentleman, though, just after our signal reports, asked how my day was going. I was a bit taken aback, as on a day as busy as this, there was little conversation. I looked up from my logbook for the first time in a while and told him my day was going wonderful and asked how his day was. He told me his day was going great, and we proceeded to have a pretty enlightening conversation about life. He gave me advice for some of my years to come, advising me to never lose sight of what I love about ham radio. To this day, I have abided. He was really nice, and I enjoyed getting to talk to him. It was a nice break from the rush of the pileup, but I was afraid I would lose it if the conversation went on too long. Remember, I was operating with my father, so I was trying to rack up contacts to beat him. About 15 minutes later, our conversation concluded, and we said 73. I said QRZ, not expecting a lot of people to still be around. However, it was like no time had passed at all, and the entire pileup was still there. It was at that moment I realized just how much I loved POTA. The people were so patient, and so kind. A 15-minute conversation with a contact that nobody had to stay around for, and pretty much all of them did. I loved it. It felt so welcoming, so inspiring, and, as weird as it sounds, it felt renewing. Hearing the blur of voices together, trying to hear just one, even just a prefix or a suffix somewhere, getting their full call sign, and finally getting to feel the gratitude of completing that contact. The conversations, the patience, the kindness in pretty much every person when telling me to have a nice day or even just 73 at the conclusion of our contact — it was rewarding.

After the activation, I had a lot of contacts to log (see **Figure 3**). Logging is generally a pretty smooth process. I like to use an app to log my contacts as I make them, so they are all already digital. However, if you like to use a logbook and write your contacts in manually, that is perfectly fine too! Back at home, I exported the file and uploaded it to the **pota.app** website. The website has become an extremely handy tool with self-service logging.

While I have been around amateur radio my entire life, I have not been around Parks on the Air that long. It was as new to me at one point as it could be to anyone, and I had to learn and grow for it to become what it has for me. POTA has provided me with so much over the years — trying out new parks, meeting new people, running into people you know on the air, getting to explore nature, building relationships with people who share the same passions as me. However, while it has been an exciting journey, I am nowhere close to finished. I cannot wait for the many more activations, many more parks, and many more contacts to come.

Figure 3, Logging contacts.

2 How to Become a POTA Hunter

By **Harold Kramer, WJ1B**

With more than 35,000 registered users on its website, Parks on the Air (POTA) continues to grow in popularity. Any operational amateur radio station can easily participate either as an activator or a hunter. An activator, as defined by POTA, "is a licensed amateur radio operator in a park on POTA's designated list [who] contacts other licensed amateurs." A hunter "is any other licensed radio operator who contacts an activator at a park."

Even though people may think of activating when they think of POTA, hunters are equally important. Both make the program a success. POTA hunters sharpen their operating skills and techniques, earn POTA awards, learn about the national and state park systems, and use their contacts to qualify for other operating awards.

To participate from your station, you must work activators who are located in state or federal parks that are listed on the POTA website (**pota.app/#/map**). POTA labels national and state parks with a single-letter four-numeral code. Parks in the United States begin with "K." For example, the POTA organization has designated Sleeping Giant State Park in Hamden, Connecticut, K-1717.

POTA operations are generally friendly and well behaved, and most POTA operators follow the DX Code of Conduct. While some activations create good-sized pileups, operations are usually well controlled and orderly. Chasing activations is less competitive than chasing rare DXCC entities. After all, a park in the USA is not Bouvet Island!

Locating Activators and Registering

To be an effective hunter, you must locate stations in state and national parks. The best resource is the POTA spotting app, **pota.app** (see **Figure 1**). Any hunter can see the activators' spots on the website, but hunters are eligible for awards and other website features only if they are registered. Registering is quick and relatively painless. Simply go to **pota.app/#/signup** and follow the directions there. POTA uses a service called Cognito to manage registrations, and the process is largely automated, with a single verification email after creating a username and password. Once you're registered with Cognito, input your call sign and name on the signup site and await approval from the POTA team. When that comes through, you're ready to hunt!

To help hunters find them, activators enter their upcoming activations on the website and spot themselves. While it looks like a conventional DX spotting network, the POTA site has advanced capabilities that let hunters control the presentation of the spotted stations. Hunters can filter activations and view by band, mode, frequency, time, park, and other data sorts. This lets hunters set up the web display based on their own operating preferences.

Logging

A hunter does not submit a log to POTA to confirm a contact. The activator submits logs directly to the POTA website and just about all activators comply with this requirement. You can view your activator-submitted contacts on the POTA website at **pota.app/#/user/logbook** as long as you are registered and signed in.

Although it's not required, a hunter should keep a log of their POTA contacts (see **Figure 2**). This is because questions sometime arise about specific contact information and by doing so, you can work toward non-POTA awards such as ARRL's Worked All States (WAS) and *CQ Amateur Radio's* WPX and County Hunters, which require confirmation from the hunter. For the same reasons, both hunters and activators usually submit their log to their preferred online logging services such as Logbook of The World (LoTW, **lotw.arrl.org/lotwuser/default**) or **QRZ.com**.

Bands and Modes

Any amateur band can be used for POTA contacts, but most POTA activators are located on 20 meters because on that band a reasonably sized antenna can be erected in a park or vehicle, and it has good propagation

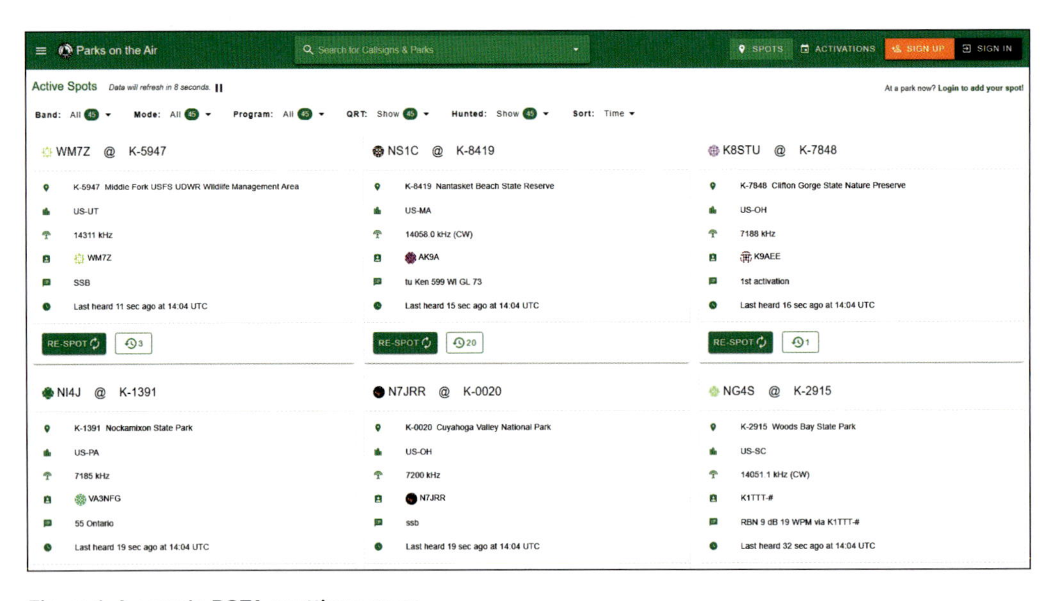

Figure 1, A sample POTA spotting screen.

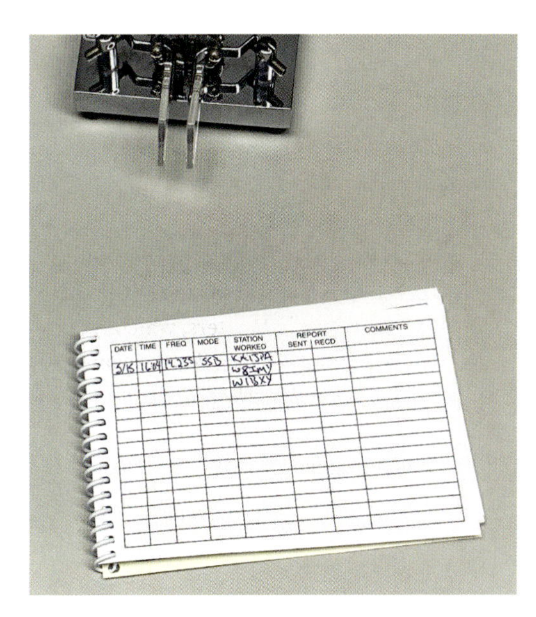

Figure 2, Pen and paper logging using ARRL's Logbook.

during daylight hours. Activators also use 40 meters during both day and night activations. With better propagation of late, there are more POTA stations on the higher frequency HF bands and on 6 meters. To attract the most hunters, activators usually operate in the General portion of the band.

POTA stations may use any mode, but most activations use SSB. A smaller percentage use Morse code. FT8 and FT4 usage continues to increase because these modes are very effective when using low RF output power, which is common for stations running on batteries. POTA operations mostly occur during daylight hours since many parks close at dusk.

Working Activators

Working a POTA activator is like working a DX or contest station but in a less hectic operating environment. However, it still requires patience, good listening skills, and persistence on the part of the hunter. Using the POTA spotting network, you will know the activator's call and frequency in advance. Once you locate an activated park, the first rule is to determine the activator's operating style. Call on or very close to the activator's frequency. Few, if any, activators work split in POTA. (*Editor's note*: Working split means calling on one frequency and listening on another.) FT8 and FT4 operate in their standard mode. FT8 operators call CQ POTA and wait for a response. I have never seen a Fox-Hound POTA activation used in the *WSJT-X* modes.

To initiate a POTA contact, simply send your call after the activator calls CQ, QRZ, or ends their previous contact, depending on their individual operating technique. There is no need to send the activator's call. Hopefully, the activator will come right back to you. Don't assume that your signal isn't getting through just because they didn't come back at first. They may be calling you back, but you may not hear them because activators often use lower-power and less-efficient antennas. Signal fading may occur as band conditions change. Don't hesitate to call a weak station since they may be copying you much stronger than you hear them!

Another reason to call again if the activator does not come back to you right away is that the park itself can be a distracting environment. Weather conditions change, curious park visitors ask questions, and park staff stop by. Activators also sometimes stop sending for a short time to change batteries or to make other adjustments, so keep listening and calling.

WJ1B's Gear

Home Station

- Icom IC-7610 transceiver
- Elecraft KPA500 amplifier
- Elecraft KAT500 automatic antenna tuner
- Icom IC-7100 transceiver for HF/VHF/UHF home operations and for some portable operations
- Heil Audio microphones and headsets
- Begali Simplex paddle
- Samlex power supplies provide DC to the radios
- Alpha Delta DX-DD dipole antenna for 80 and 40 meters at about 50 feet
- K4KIO Hex Beam antenna on a 40-foot Rohn 25 tower for 20 – 6 Meters
- Yaesu G-800SA rotator
- Diamond X30 antenna up 25 feet for 2 meters and 70 centimeters
- *DX Lab Suite* logging software
- *N1MM Logger* for contests
- Logs uploaded to Logbook of The World (LoTW) and to **QRZ.com**
- *WSJT-X* for FT4, FT8, and MSK144
- *JTAlert* for visual alerts
- HP PC with Windows 10 and dual monitors

Portable and Field Operations

- Icom IC-7100 transceiver (used occasionally and mentioned above)
- LDG tuner
- Icom IC-7300 transceiver
- 12-amp Bioenno battery
- Powerwerx Battery Box
- 5-amp Buddipole battery as backup
- Buddipole antenna for HF, used in a vertical configuration
- *HAMRS* portable logging or pen and paper

Public Service and Emergency Communications

- Kenwood TH-F6A VHF/UHF triband handheld transceiver
- Accessories such as headsets, speaker microphones, battery packs
- 2-meter and 70-centimeter portable antennas
- AnyTone AT-D878UV PLUS for portable and DMR

Once the activator acknowledges your call, you should know there is no required exchange. Any legal contact between an activator and a hunter counts, as long as both operators follow the other POTA rules. Most contacts include accurate signal reports and state or DX location, though again, it's not required. You can also thank the activator for being out there! Generally, POTA contacts don't entail a lot of conversation. If the activator isn't busy, they may exchange other information about the park or their station (see **Figure 3**). Hunters get credit for working a park only once on a given band on any given day. So, you may want to keep track of your contacts to make sure you aren't working a duplicate.

If you are on SSB, use standard phonetics. Make sure that you copy the activator's call sign, not the call sign of another hunter. On CW, POTA contacts are sent from 13 to 22 wpm at the most, and it's best to match your CW speed to speed of the activator, though sometimes park environment make receiving more difficult, so you might even consider sending a little slower than the activator. Send your state as its two-letter abbreviation, e.g., "CT" for Connecticut on CW or "Charlie Tango" on SSB.

Activators earn special awards for working other parks. This is called a park-to-park contact. You may hear stations calling "park to park." These contacts can be difficult to complete, and it is discourteous to transmit when an activator in one park tries to work another. As a final courtesy, hunters should spot or re-spot the activator on the POTA website. This is very helpful for both activators and hunters.

Awards for Hunters

POTA offers a variety of awards for all types of on-air operating activity. There really is something for everyone. While top hunters have worked and confirmed over 10,000 parks and 45,000 contacts, a hunter can obtain an award after confirming only 10 parks and for confirming unique DX stations in increments of five. A Worked All States Award is also available. Hunters qualify for these awards with confirmed POTA contacts using any combination of modes or bands.

There are also awards for working parks at specific times of day. POTA gives out time-based awards for 100 parks confirmed in the

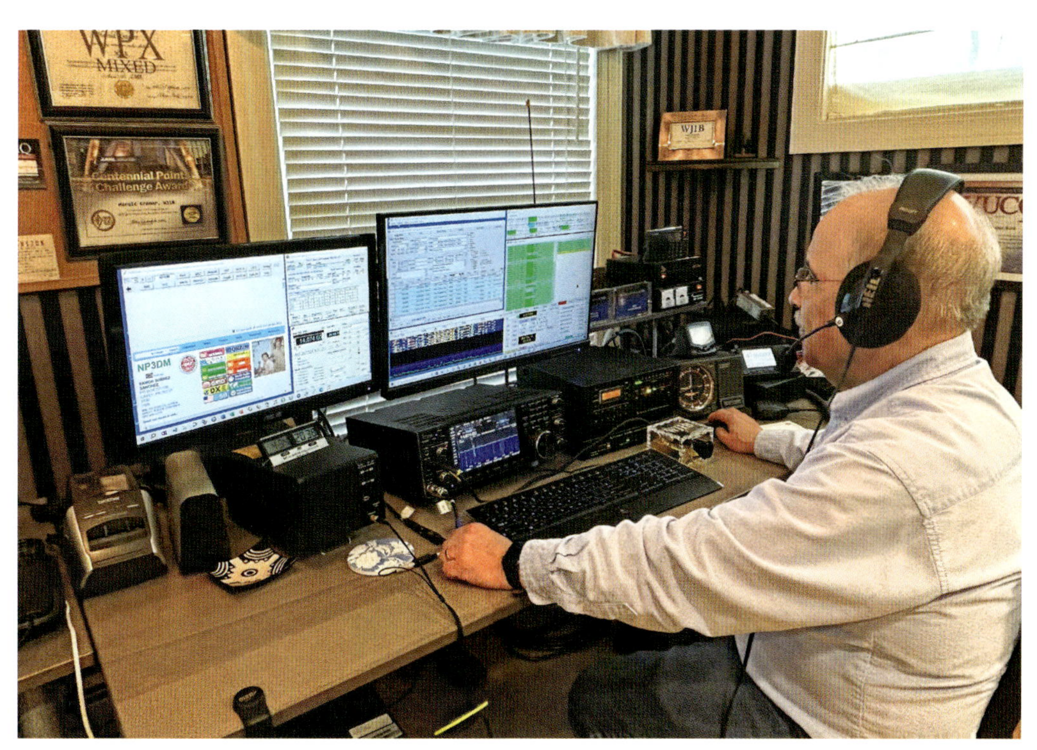

Figure 3, WJ1B operating at his home station.

six-hour period beginning at 2:00 UTC for the "Early Shift" award while you can earn "Late Shift" awards for working 100 parks during the eight-hour period beginning at 18:00 UTC. Because the sun rises and sets at different times depending on where you are located, the exact times for these awards vary with a station's longitude, so check the time zone map on the POTA website for the correct start and end times for your location.

There is no need to apply for a POTA operating award. Once you are registered on the POTA website, it tracks and awards operating certificates automatically. When you achieve an award, the website displays a professionally designed, personalized, PDF certificate that you can download and print (see **Figure 4**). The website has extensive award tracking information so you can easily monitor your progress.

POTA also sponsors operating events during the year. One of my favorites is the Support Your Park Weekend. These occur each season during the third full weekend of the month in January, April, July, and October. These weekend events are designed to generate some fun and increase operating activity by getting more activators in the parks and having more hunters on the air to work them. There are other events including a New Year's Week event and an Annual Plaque event. Details can be found on the POTA website at **pota.app/#/events**.

If you would like to learn more, check out the many videos and documents on the web about POTA. Give it a try, have some fun, and enjoy hunting POTA!

I would like to thank: Dave, NZ1J; Shawn, KC1NQE; Eric, KB1JL; and Bart, N1BRL, for sharing their knowledge and expertise about POTA with me for this article.

Figure 4, A sample hunter award.

3 CW and POTA

An Introduction

By **Bill Brown, K4NYM**

I arrive at the Marcus Road Trailhead, part of the Hilochee Wildlife Management Area (WMA), K-6309, and park under the shade of the tall pine trees (see **Figure 1**). The trailhead is just off CR 474 in rural Lake County, Florida, about 10 miles from Disney World, as the crow flies. The shade is important on this hot and humid summer morning. The air is still and disturbed only by occasional dump trucks hauling sand from the nearby slurry mines. Since the temperature has already reached the mid 80s and the forecast predicts high 90s, with thunderstorms in the afternoon, I decide against venturing down the trail and instead activate the park from the cozy confines of my "POTA mobile." That's what I call my 2017 Toyota Camry. I set up my Kenwood TS-480SAT and my N3ZN Ironman single-lever paddle on an old second-hand portable desk that I place on the front passenger seat. I then deploy my 17-foot stainless-steel whip on the tri-magnet mount on the trunk of my car. After finding an open frequency on 20 meters and announcing my activation on the POTA spotting page, I start calling CQ. Much to my surprise, my first two contacts pin my S-meter, meaning they're either putting out a lot of power or they're close by. As it turns out, they're close. One contact is set

Figure 1, A trailhead at Hilochee Wildlife Management Area.

up at a trailhead in a different section of the park. The second contact is a few miles up the road at a third trailhead, also at the Hilochee WMA. I have just made two park-to-park contacts with two different activators all in the same park, and none of us knew the others were going to be there. POTA's popularity and the explosive growth of CW operating is what made such a unique experience possible.

Not long ago, many CW operators thought the elimination of the Morse code licensing requirement would be the death of the CW mode. Not only has that not been the case, but, in fact, the CW mode is experiencing a resurgence that can, in no small part, be attributed to portable work, and specifically, operating activities like Parks on the Air (POTA) and Summits on the Air (SOTA). POTA not only attracts no-code hams to CW but has also kindled renewed interest from former CW operators.

The Uniqueness of CW in POTA

Phone is the dominant mode in POTA, specifically SSB. While there is skill involved in knowing how to get through to an activator and complete a contact, phone doesn't require any special training. You simply pick up the microphone and speak. On the other hand, CW requires proficiency in a different language — a binary language — with its own distinct alphabet. Morse code contains its own idioms, colloquialisms, and even grammar. A trained ear can make out the "personality" of the sender.

But just as a child's vocabulary starts with simple words and slowly expands, so it is with the beginning CW operator. It isn't necessary to have years of experience to enjoy CW, and POTA provides an opportunity to gain that experience and have fun in the process.

CW activating is also unique in that an activator can fit everything necessary into a bag no larger than a shaving kit. A pocket-sized transceiver, a wire antenna, and a portable key or paddle are all you need to get on the air.

CW's Popularity in POTA

How popular is CW activating in POTA? Consider the growth in both total CW activators and CW contacts. In POTA's first year, 2017, there were 97 total activators and 37,500 CW contacts. **Table 1** shows the total number of POTA CW activators and their contacts in each of the years since 2017. You can see that CW activity is growing in POTA. Matt Heere, N3NWV, one of POTA's administrators, provided me with the information in Table 1. (His essay titled "Big Benefits from Small Stations" appears later in this book.)

The statistics for the current year (2023) are not complete, but at the end of the first

Table 1
Number of POTA CW Activators and Total Contacts by Year

Year	*Number of Activators*	*Number of Contacts*
2017	97	37,500
2018	172	69,064
2019	252	79,425
2020	422	160,761
2021	956	422,264
2022	1,612	811,142
2023*	1,447	494,679

*Numbers are for Quarter 1, January – March, 2023.

quarter there have been 1,447 activators and 494,679 contacts. If this trend continues the current year may finish with over 2,000 activators and 1.9 million contacts! That is impressive growth.

Many factors contribute to this growth, and social media is an important one. Social media provides a way for CW activators to share the fun they are having with "the original digital" mode. Twitter (**twitter.com/POTAspots**), Slack (**pota.app/slack**), and POTA's Facebook group page (**facebook.com/groups/parksontheair**) provide a forum for individuals to share their personal journey with CW. Members post pictures, videos, and comments from hams making their first CW activation. These posts often receive many likes and supportive comments, which helps interest other hams, who may then decide to give CW activating a try.

Activating Using CW

Challenges in CW Activating

Operating portable CW differs from operating at home (see **Figure 2**). There are certain challenges activators using any mode regularly face: rain, lightning, heat, cold, insects, and equipment failures. Even unexpected success presents a challenge. CW hunters are persistent, and it is not surprising for a CW activator to instigate a DXpedition-worthy pileup. When dozens of stations are simultaneously trying to contact you, picking out a call sign and managing a pileup can be challenging. Many new activators develop this skill on the fly.

Send and receive speed can also present challenges for new activators. My CW

Figure 2, Bill Brown, K4NYM, operating portable CW.

experience goes back to my time as a Morse intercept operator in the United States Air Force. I was able to head copy at 30 words per minute, but that skill took time and practice to develop. An activator proficient at 13 words per minute may struggle with a hunter sending 20 words per minute and vice versa. Thankfully, POTA has plenty of courteous activators and hunters who will slow down (QRS) to match the speed of the slower CW operator.

Advantages of CW Activating

CW is a narrow-bandwidth mode. It doesn't require much power. Five watts and just fair propagation are all you need to make the requisite 10 contacts and activate a park. CW operators have an advantage over those relying on SSB, with its wider bandwidth and its susceptibility to noise.

Having the ability to operate CW gives an activator another arrow in the quiver. This is especially helpful during SSB contests. When the phone portion of the bands are full of contest stations, the CW operator can easily switch modes. This ability can sometimes be just what is needed to get those 10 contacts. CW also offers an additional band — 30 meters — that isn't available for phone use and that generally doesn't get busy during contests because of its status as a WARC band.

CW operators are a tight-knit group. The famous *Field of Dreams* line, "If you build it, they will come," applies to the relationship between CW activators and hunters. If you call "CQ POTA," and propagation is in your favor, don't be surprised if you find yourself in the thick of a pileup.

K4NYM's Gear

- Icom IC-7100 transceiver
- Begali Traveler paddle
- Radiosport CW headset
- Microsoft Surface Pro notebook computer
- *HAMRS* logging software
- Gabil GRA-7350T antenna
- Diamond K400 trunk/hatchback mount

Types of Equipment

As mentioned earlier, it doesn't take a lot of equipment to activate CW in POTA (see **Figure** 3). Many CW activators have "go-bags" that are minimalist in nature (see Thomas Witherspoon, K4SWL's, essay, "The Art of the Self-Sufficient QRP Field Kit" and Matt Heere, N3NWV's, essay, "Big Benefits from Small Stations" later in this book). A small pocket-sized transceiver, a low power end-fed half-wave (EFHW) antenna, and a portable key or paddle easily fit into a backpack. Then again, an activator can set up at a picnic table with a more robust station that rivals the one in their home.

The list of portable, low power, "go-bag"-size transceivers is extensive and includes popular models like the Elecraft KX2/KX3, Mountain Topper line, QCX Mini, and the Xiegu X5105. Frequently used shack-in-the-box, 100-watt rigs include the Yaesu FT-891 and Icom IC-7300.

Antennas run the gamut from hamsticks, EFHWs, telescoping whips, and random wires to coil-loaded antennas such as those produced by Wolf River Coils and Super Antenna. Activators also favor magnetic loop antennas like those from Chameleon, Alpha, and AlexLoop.

You can't have a CW activation without a key or paddle. Myriad manufacturers and styles can be found in the parks. A short selection includes BaMaTech, CW Morse, Kent, Begali, and N3ZN. And let's not forget home-brew keys and paddles! Each has its advantages, and the activators who use them can be fiercely loyal.

Morse Code Proficiency

Whether it is a POTA contact or not, the first time someone sends CQ can be quite intimidating. Many fears and apprehensions bubble in the new CW operator's mind. "Will I send well enough?" "Will I copy accurately?" "What if the other station sends too fast?" "Will I use the right format?" If those questions are going through your head, don't worry. Every CW operator asked the same ones when starting out. The good news is that you don't have to be a CW veteran to make a POTA contact. If you possess a working knowledge of Morse code

Figure 3, A look at Bill's portable operating setup.

and can handle a repeatable contact formula, you will do fine in POTA.

To make any task easier to learn, it can help to break it down into short, easy, and repeatable bits. POTA contacts meet these criteria. Contacts are typically short: usually nothing more than an acknowledgement, a signal report, and a valediction. The exchange is easy to send and copy because it's generally predictable. And the form itself is repeatable for each new contact. Here is an example of a common POTA CW contact:

CQ CQ POTA DE K4NYM K4NYM K
WDØACB
WDØACB DE K4NYM GM UR 599 599 BK
K4NYM DE WDØACB GM UR 599 599 KS BK
WDØACB DE K4NYM R TU 73 DIT DIT
WDØACB DIT DIT

Basically, the contact is an exchange of call signs (K4NYM, WDØACB) and signal reports (599). Sometimes hunters will include their state or province. DX stations typically just send their call sign and signal report. (See **Table 2** for a quick guide to the Readability,

Table 2
The RST System

Readability

1 Unreadable.
2 Barely readable, occasional words distinguishable.
3 Readable with considerable difficulty.
4 Readable with practically no difficulty.
5 Perfectly readable.

Signal Strength

1 Faint signals, barely perceptible.
2 Very weak signals.
3 Weak signals.
4 Fair signals.
5 Fairly good signals.
6 Good signals.
7 Moderately strong signals.
8 Strong signals.
9 Extremely strong signals.

Tone

1 60-cycle ac or less, very rough and broad.
2 very rough ac, very harsh and broad.
3 Rough ac tone, rectified but not filtered.
4 Rough note, some trace of filtering.
5 Filtered rectified ac but strongly ripple-modulated.
6 Filtered tone, definite trace of ripple modulation
7 Near pure tone, trace of ripple modulation.
8 Near perfect tone, slight trace of ripple modulation.
9 Perfect tone, no trace of ripple modulation of any kind.

If the signal has the characteristic steadiness of crystal control, add the letter X to the RST report. If there is a chirp, add the letter C. Similarly, for a click add K. (See FCC Regulations §97.307, Emissions Standards.) The above reporting system is used on both CW and voice; leave out the "tone" report on voice.

Strength, and Tone (RST) system.) It is not even necessary to identify park-to-park contacts, since POTA's website matches logs once they are uploaded, and it credits park-to-parks automatically.

Starting Out

Every POTA enthusiast, like pretty much every ham, has a story about how they got started in the hobby. Jim Williams, N4JAW, shared the story of how he got started in POTA.

Jim was first licensed in 1962. Code was a license requirement at that time, but he didn't invest too much time in CW in the beginning. He enjoyed a long career in radio and television broadcasting, and, after retiring, moved into an apartment which made getting on the air difficult. While he dabbled off and on with the hobby, the COVID pandemic renewed his interest, especially in CW. Jim has many hobbies and passions, but he took a special interest in POTA. In August of 2020 he first activated Beargrass Creek State Nature Preserve, K-7956, which is close to his home in the Louisville, KY area. Beargrass Creek is an urban green space that is popular with bird watchers and hikers. With the park being so close to his home, it was relatively easy for him to work in an activation, and when 2021 rolled around Jim was activating Beargrass Creek multiple times a week (see **Figure 4**). CW was at the center of Jim's activations. Instead of a car, he often rode his bike to the park. He has become a regular fixture and is often approached by interested people

Figure 4, Jim Williams, N4JAW, activating Beargrass Creek State Nature Preserve. (Photo courtesy of Jim Williams, N4JAW)

wanting to know what he is doing. Jim gladly assumes the role of POTA, CW, and amateur radio ambassador. He regularly posts videos of his serendipitous encounters with visitors on his Twitter account **@N4JAW**. Jim enjoys low power CW with his QCX Mini and HB1B transceivers. He typically uses a wire antenna supported by a 7-meter fishing pole. A Radio Adventure Gear paddle allows his fingers to do the walking. Jim told me that POTA brings him back to his childhood and those early days when he first got into the hobby. Like many of us, he fell in love again with the wonder of it all, and POTA was there to help him do that. His advice to aspiring POTA CW operators is to "Get on the air!" In Jim's case, perhaps the words of advice Bilbo Baggins gave his nephew Frodo in J.R.R. Tolkien's *Lord of the Rings* would be fitting, "It's a dangerous business, Frodo, going out your door. You step onto the road, and if you don't keep your feet, there's no knowing where you might be swept off to." When Jim experienced his first POTA activation, he was swept into the world of CW, and he's never turned back.

Jim's advice is the advice I give to anyone thinking about giving CW and POTA a try. Get on the air! The only way to find out if this mode is for you is to use it. If you do, don't be surprised to find yourself swept away as well — into a thrilling POTA CW community.

PART 2
Get on the Air

Harkness Memorial State Park, K1677 — New London County, CT

Photo by Joe Buglewicz

4 13 Parks in 4 Days in Rural Midcoast Maine

A POTA Rove

By **Kevin Thomas, W1DED**

In April 2022 I reentered the hobby. Combining ham radio with the Maine outdoors seemed like the perfect pairing. I threw myself into learning how to activate Parks on the Air (POTA) parks and got hooked in the process. I especially enjoyed the thrill of locating "new to me" parks, resolving the setup challenges at each, and making as many contacts as possible. It didn't take long to activate most of the parks near my home in Maine. But if I was going to activate all 142 eligible Maine POTA parks — as I hoped to do — I'd need to travel, find accommodations, and activate parks in a cluster. That's how I found myself in rural, midcoast Maine, over 100 miles from home, trying to activate 10 parks in 3 short days.

Preparation

Getting to activating took some planning. In order to visualize a possible route, I built a custom Google map to include all the Maine parks I'd activated to date, the parks I hadn't gone to, and a special note in the event that no one had been there yet. I added coordinates for entry points along with website URLs, photos, and notes. A scan of my completed map revealed park clusters where 10 activations might be tenable, with an obvious cluster centered on Stockton Springs, Maine.

The route I laid out would be critical to minimizing zigzagging travel miles — a feature that I knew would make activating Maine's rural countryside difficult. I estimated travel times to and from parks as well as the distance to my cabin rental. After playing with variations, all impacted by the various parks' open/close times and accessibility issues, I came up with an efficient itinerary. I figured that a well laid plan would exponentially improve my odds of success.

My destination and route decided, I turned to creating an equipment list that would be versatile enough for a range of conditions with the redundancy required to keep me operating should some component fail when I was far from home. I started with my usual go-bag, packed several antennas, batteries of various sizes, a power supply, and then various accessories to allow for more comfortable operation (see **Figures 1** and **2**). For the full list, see the sidebar, "W1DED's Gear."

W1DED's Gear

Transceiver and Power

- Kenwood TS-480-HX transceiver
- MFJ antenna tuner
- SWR meter
- Heil BM-17 headset
- Bioenno lithium batteries (various sizes)
- MFJ 75-amp power supply

Antennas

- Two Buddipoles, one of them the long version, with tripods and masts
- Chameleon CHA LEFS8010 (Lightweight End-Fed Sloper) end-fed half-wave

Other Necessities

- Folding camp table
- Folding chair
- Toolbox
- Several small gear bags
- Tent
- Portable propane heater
- Portable ice shack (for the chilly Maine nights)

Figure 1, All the gear ready to pack.

Figure 2, Everything loaded.

Day One

Day one began at sunrise and my car was packed by 8:00 am. My plan was to head north to the James Dorso Wildlife Management Area (WMA), K-8455, then on to the Frye Mountain Wildlife Management Area, K-8290, followed by the Hurds Pond WMA, K-8470, before I finished at the easily accessible Swan Lake State Park. I'd then check into my rental cabin, regroup, and drive the 10 minutes to Fort Point State Park, K-2389, for a Late Shift activation.

A Brief Detour

It was to be a busy but productive day — except nothing went as planned, starting with my departure time.

In what felt like a very "Maine" set of circumstances, just as I was readying to leave, a local trapper contacted me about removing a family of skunks that had taken up residence under my deck and had been bothering my dog. He wanted to come over right away to start removing the skunks. When I weighed the costs and benefits of putting off my departure, I had to come down on the side of delaying. In total, waiting for the trapper put me on the road a full three hours late. However, I came home to a skunk-free house, a trade-off I considered well worth it.

First Activations

My first park, the James Dorso Wildlife Management Area, encircled the Bartlett Stream Reservoir in Searsmont, Maine (see **Figure 3**). Despite my late start, the access was exactly what I had hoped for, just off Route 3. I couldn't believe that this park hadn't been activated before and felt rewarded to get the all-time new one (ATNO) with my call sign permanently listed as the first-time activator. I don't love parking lot operations, but no one was there, and the view out across the pond was beautiful. It was threatening to rain, so I made the decision to operate from my car. The Buddipole went up in the usual vertical position that includes a nine-and-a-half-foot whip with four 22-inch arms raised to about eight feet. Four wires, each approximately 17-feet long, extended from the Versatee. Without any adjustment, the SWR was near perfect, and people started responding to my CQ POTA immediately at 1825 UTC. I logged 125 contacts within 54 minutes.

Next stop was the Frye Mountain Wildlife Management Area. I didn't have time to hike

Figure 3, View of the James Dorso (Ruffingham Meadow) Wildlife Management Area, K-8455.

up Frye Mountain, but my research indicated a maintenance garage within the boundary area, which I found pretty easily. The familiar government-issued, brown wildlife management area sign hung nearby. Given the misty weather, I set up the same way I had at Dorso and operated from my car. I logged 103 contacts and scored a second ATNO.

Since I was running late, I deviated from my plan and skipped Hurd's Pond WMA. The access point was never clear, and I just didn't have the time to figure it out. Swan Lake State Park was only 30 minutes away, and I assumed it would be a quick activation (state parks usually are). Instead, I was met with a closed gate, a long entry road, and no idea how far the walk-in would be. I abandoned my plan, again, continuing the day's trend. I kept driving to my cabin rental where I regrouped before my planned Late Shift activation at nearby Fort Point State Park.

The accommodations at Steamboat Wharf Cabins were perfect for my three-night stay. I moved in with some of the gear and set up the power supply to recharge my battery before heading off to Late Shift. Once again, I faced a closed gate. This time it was clear the distance into the park was more than I wanted to tackle, especially for a Late Shift. Feeling defeated, I called it a day and headed back to the cabin to upload my logs.

I use the *HAMRS* logging software to record my POTA contacts. It's simple and requires only the basics. When POTA pileups are happening, I've found that's all I have time for. Sometimes when I have cell service, I use my phone as a hotspot to get notified of POTA spots and to use the QRZ call-sign check (**qrz.com**). In general, I prefer to stay up to date with my logs so hunters get quick credit for their contacts.

Back at Steamboat Wharf Cabins, my end-of-day process was to do a cursory log review, export the ADIF file, and then upload the contacts immediately to the POTA site.

Day Two

Day two started with a 20-minute drive to Fort Knox State Historic Site, K-8291 (see **Figure 4**). This park sat dramatically on the edge of the Penobscot River, overlooking the town of Bucksport and the stunning Penobscot Narrows Bridge. Assessing my options, I realized the best setup location was up the hill near the fort, only a short walk from the parking lot. It was the highest point in the park, with no power lines nearby and I'd be out of the way of visitors. I was only the second person to activate the park and made 106 contacts in under an hour.

Next up was the Howard L. Mendall WMA, K-8461 (see **Figure 5**). I was learning WMAs were wildcards. The access points were often obscure and this one was no different. On the

Figure 4, W1DED operating at the Fort Knox State Historic Site, K-8291.

Figure 5, W1DED's setup at the Howard L. Mendall (Marsh Stream) Wildlife Management Area, K-8461.

west side of the South Branch River, I found a heavily rutted dirt road and a barely visible brown WMA sign. The road continued far into the waterway taking me literally to the middle of the river. This was a spectacular site: especially so during Maine's legendary fall foliage season. Brisk winds across the waterway made antenna setup and operation a bit challenging. I was on the air by 1639 UTC and made 81 contacts in about 50 minutes from this very memorable park location.

Moose Point State Park, K-2396, was in Searsport and only 20 minutes away, but first I wanted to scope out the Sandy Point WMA, K-8466. This one was close to my cabin rental, and I thought it might work for a Late Shift activation. The reconnaissance trip was well worth it. I found it to be an accessible WMA and I made it my intention to do a Late Shift sometime within the week.

I arrived at Moose Point only to learn that they close their gates promptly at 5:00 pm versus the expected sunset rule. My activation needed to be quick. The good news was that a parking lot sits adjacent to a large open field filled with picnic tables at the edge of the Atlantic. I got on the air in record time after a quick SWR check. The band was hot and I immediately regretted that I didn't have much time. I logged 103 contacts in 40 minutes, and I pushed it as long as I could before driving out of the park with only two minutes to spare.

After stopping by my cabin to grab a fresh battery, I tried Fort Point State Park again (see **Figures 6** and **7**). I discovered an easier way into the park that bypasses the gated entrance. Lighthouse Road took me to a small parking lot within a stone's throw of the lighthouse and prime setup area. But the sign said, "Closed at Sunset," which would not work for my Late Shift plans. As I considered a workaround, a pickup truck drove past and I thought, "Could this be the park ranger?" Sure enough, he was and I gave him a brief summary of ham radio and Parks on the Air and asked his permission to set up and stay well past sunset. He kicked the dirt, clearly thinking it over before acquiescing. With that, I decided to go for broke on this activation. I set up both antennas, one for 20 meters and another for 40, so I could easily switch over when the time came. It was forecasted to be a cold night, so I pulled out the portable ice shack and set up the propane heater. This all took a lot of time. I started operating at 2300 Zulu and it was busy — 127 contacts within an hour. I didn't want to get stuck in a 20-meter pileup during the transition

Figure 6, Sunset at Fort Point State Park, K-2389.

Figure 7, Late Shift activation at Fort Point State Park, K-2389.

to the new UTC day, so I made the coax switch and hopped over to 40. I was making contacts in no time, but it was a surprising slog. It took me nearly an hour and 15 minutes and lots of CQing to get to 88 contacts. This was not what I had hoped for given the investment of setup time, so I decided to call it quits. I knew I still had at least half an hour ahead of me to pull down both antennas and to pack all the gear, including the ice shack, in my car. It was a very late night by the time I was in the cabin, but uploading logs for five park activations was more than enough to offset any exhaustion.

Day Three

Holbrook Island Sanctuary State Park, K-2391, started day three (see **Figure 8**). This one was a 50-minute drive from my basecamp. When I arrived at Holbrook and the dirt road to my setup location on Indian Bar, it was clear that I had stumbled upon something very special. Holbrook consisted of over 1,300 undisturbed acres that clearly got very few visitors. I felt blessed to have been drawn here by ham radio. Once at Indian Bar, I found a chance to set up at the water's edge to take advantage of the saltwater boost. This 20-meter activation was going wonderfully until I noticed the tide was coming in much faster than I had expected. As the tide started to overtake my position, I broke down and headed for high ground. Lesson learned: don't get greedy. Stay above the high tide mark. I still managed 70 contacts in my 30 minutes on the air.

Swan Lake State Park, the one that thwarted me on day one, was next on my list. I decided that if they were truly closed for the season, I'd just hike in to get the activation check mark. I found an empty, pristine rock beach with a lifeguard station that would be my operating perch. By 1746 UTC I was operational and made 69 contacts over the next 50 minutes.

I'd been dreading Hurds Pond since day one. Gaining access to this WMA was absolutely unclear. No one had yet activated it, and I couldn't find any clues online. So, I just started driving and hoping for the best — maybe a chance encounter with someone who lived nearby. And that's exactly what happened. I stopped to ask a neighbor who was out checking her mailbox. She pointed off to the north and said she wasn't sure if it was an official road and that it had recently been gated, but it was the road fishermen used. I followed her directions and parked in a small clearing in front of a gate. A sketchy trail led into the WMA. I packed my gear and hiked in. Eventually, I came upon a wooden shelter alongside a clearing and decided this was likely as good as it would get. I set up on 20 meters and operated for just about an hour before hiking out, having logged 118 contacts.

Figure 8, View of the beach at Holbrook Island Sanctuary State Park, K-2391.

Sandy Point WMA, which I had scouted earlier, was next up. This was a spooky, desolate place at dusk and I decided operating from my car would be best, so I could get out quickly when I was done. Given the tough time I had at Fort Point, all I hoped for was the needed 10 contacts, but the pre-Late Shift pace was

W1DED on YouTube

To see Kevin's interviews with a variety of hams, check out "W1DED WW Ham Radio" at **youtube.com/@w1dedworldwidehamradio**.

rapid, and I quickly logged 146 contacts on 20 meters. I was ready to move to 40 for Late Shift and found myself unprepared and fumbling around in the dark to add a 40-meter coil to my Buddipole, extend the radials, and check the SWR. I've often struggled tuning for 40 meters and this night was no different, except now I was trying to adjust things under headlamp light as the clock ticked. I finally got the SWR to an acceptable level and was back on the air. Forty meters was hot, and I racked up 176 contacts. Surprisingly, Sandy Point WMA would turn out to be my best park, with 322 total contacts in just over two hours of operating time.

Bonus Parks

On day four, I headed home. The trip was a huge success, and I achieved my 10-park activation goal. Never content, I also realized that I would pass close to a few parks on my way south. Before long, I pulled into my first "bonus" park — Camden Hills State Park, K-2384. I followed that with Owls Head State Park, K-2399, and Birch Point Beach State Park, K-2382, bringing my trip total to 13 unique park activations, 15 total activations, with over 1,600 contacts (see **Figure 9**.)

Figure 9, W1DED's activating position at Birch Point Beach State Park, K-2382.

5 Advantages of Wire Antennas

Vertical Dipoles, Vertical Random Wires, and Nonresonant Doublets

By **John Ford, ABØO**

Wire antennas have been a staple of ham radio for almost a century because they provide high radiation efficiency with low cost, flexible designs, and easy deployment. While a compact coil-loaded short whip may seem convenient and user-friendly, it seldom delivers much efficiency, so little that as much as half or three quarters of the transmitter power is lost as heat in the antenna system and not radiated into the ionosphere.

Examination of the Humble Dipole

It's no mystery why most multielement beam antennas are built around a basic dipole. It's because the dipole is very efficient when properly constructed.

So, if we look at the dipole in detail, we see that there are a few notable characteristics:

- The radiation resistance or impedance is very close to the impedance of the coaxial cable that typically feeds it.
- The dipole is typically resonant at or close to the desired/design frequency when properly sized.

Wire Antennas and Parks

Whether the park website says so or not, almost all parks and/or states have environmental management laws that prohibit digging, harvesting, cutting, pruning, breaking, mowing, etc. The reason for these laws is to prevent non-native organisms from spreading to other parks or states.

In the context of POTA, a stake in the ground may be considered "digging," and a line in a tree may be considered "breaking or pruning," and in both cases, soil or vegetation from this park can possibly be carried to another park in your vehicle, not to mention that you disturbed or damaged the park premises by doing what you did. My golden rule of POTA is, "No stakes in ground, and no wires in trees!"

It takes some thought to design a wire antenna system you can set up in the middle of a concrete parking lot, but many of us have been doing this for years. It can be done.

• There is usually no need for a matching device when the antenna is resonant.
• If the dipole is properly elevated, ground return losses are low because the dipole provides its own return path for the RF energy.

Sparing you the math, all the above computes to an efficient antenna. Therefore, if you have 10 watts at the antenna feed point, then pretty close to 10 watts of RF energy will be radiated by the antenna toward the ionosphere.

Uncommon, but Very Practical Wire Antennas for POTA

I won't discuss traditional wire antennas and dipoles. There have already been thousands of pages written on those. The antenna variations that aren't typically the subject of many publications, if any, are as follows:

- Vertical dipoles
- Vertical random wires
- Nonresonant flat tops or doublets

These options offer advantages that can allow you to have a smaller footprint and maintain antenna efficiency. Note that I typically use common #18 or #22 insulated stranded wire for my POTA antennas, but any insulated stranded wire up to #14 can be used.

I propose two of these antenna types for activating, and one type for hunting.

Vertical Dipoles (Activator's Choice)

The vertical dipole is electrically identical to the simple horizontal dipole, except that it is mounted vertically, either suspended from a high support, or attached to a nonconductive mast. (A close relative would be a "sloping dipole.") If the mast is tall enough, the entire dipole can be vertically mounted, and the coax can be run horizontally to an elevated operating platform. If the coax comes away from the mast for at least ¼ wave before sagging to the ground, the antenna won't be affected by the coaxial cable's proximity.

A more practical configuration that doesn't give up efficiency (and doesn't change tuning, usually), is the "upright L" configuration.

This is my preferred configuration for activating, because the efficiency of the dipole is preserved, but in a shorter height and a smaller footprint. As you can see in **Figure 1**, the *EZNEC* antenna plot shows a maximum lobe at about 20 degrees, which compares favorably to most vertical antennas, and is somewhat better than an inverted-V or horizontal dipole. On the *EZNEC* plot, the upright-L shows a slight directivity towards the horizontal leg, but during several hundred POTA activations with this antenna, I have not found any practical evidence of any directivity.

The upright-L, when mounted on a fiberglass mast, has one vertical leg attached to the mast (with simple painter's tape) and the other leg strung horizontally to a second,

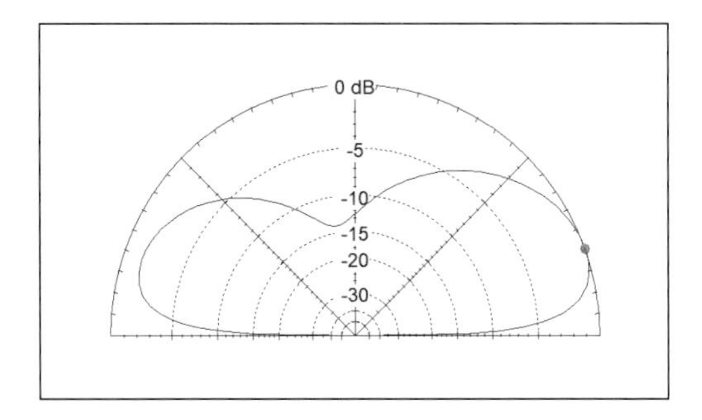

Figure 1, *EZNEC* plot of the upright-L's radiation pattern.

shorter mast or support. The horizontal leg is not a radial and it is in no way connected to ground. It is simply the second leg of a regular dipole. See **Figure 2**.

The important detail about the upright-L is that the vertical leg is connected to the coax *center conductor,* and the horizontal leg is connected to the coax *braid*. In the absence

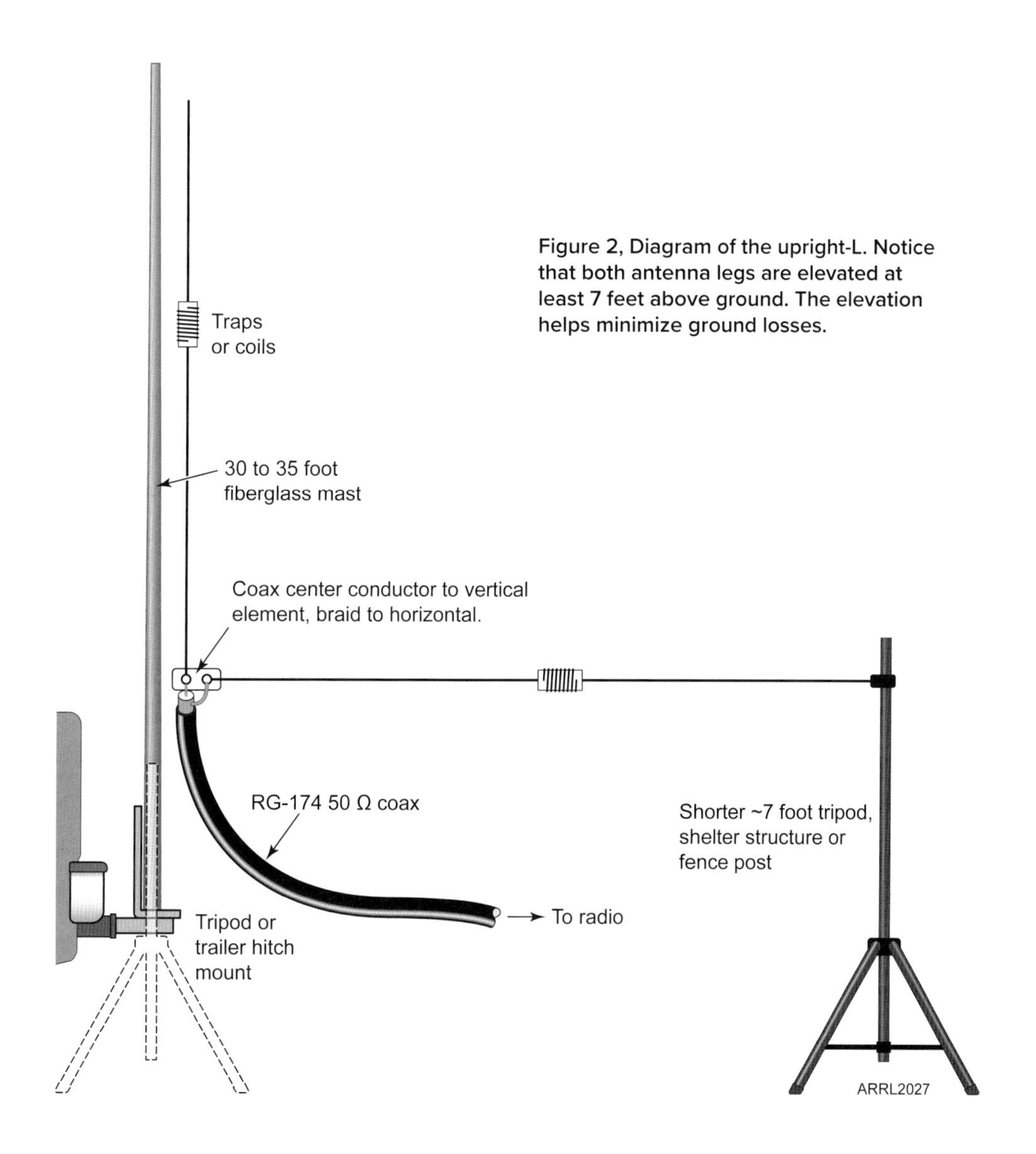

Figure 2, Diagram of the upright-L. Notice that both antenna legs are elevated at least 7 feet above ground. The elevation helps minimize ground losses.

of a balun at the feed point, the coax center conductor has most of the RF current, so the vertical leg is "working the hardest." The other/horizontal leg, as with any dipole, provides a return path for the signal. Remember that the wire dimensions for the upright-L are the same as for any regular dipole as shown in many antenna books. The only difference is that the dipole is mounted in an L shape, rather than as a horizontal dipole or

ABØO's Gear List

My personal POTA kit that I have used for more than 700 park, summit, or island activations is shown in the image below, and the itemized list is as follows:

- Elecraft KX3 transceiver with stock microphone
- Bioenno 6 amp-hour LiFePO4 battery
- Regular stereo earbuds
- SOTABeams TravelMast 32 feet
- Pacific Antennas 20/40 trap dipole
- Elecraft bolt-on CW paddles
- 25-foot RG-174 BNC coaxial cable
- ARRL MiniLog

Not shown:

- 3M yellow plastic painter's tape
- Leatherman multitool
- Three 10-foot lengths of paracord
- Inexpensive regular backpack with pockets

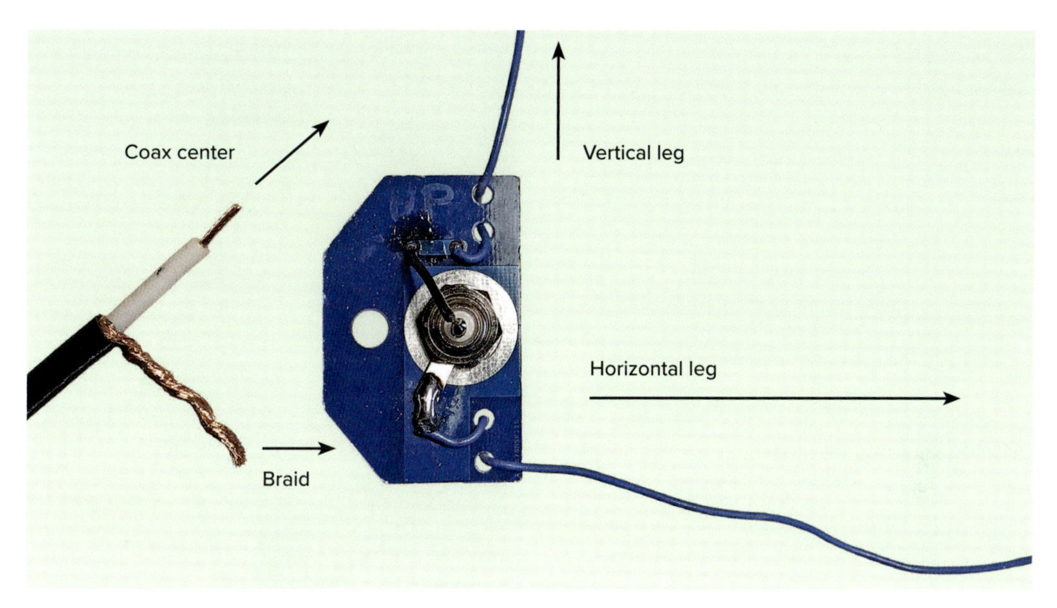

Figure 3, Feed point connections for the upright L.

Figure 4, The upright L in use. Notice the 20-meter traps in the 20/40-meter dipole.

inverted-V. For the feed point connections, see **Figure 3**.

For my POTA activations, I use a 20/40-meter trap dipole that is available as a kit, as shown in **Figure 4**. Even though the 40-meter dimension is shortened by the trap coil, the antenna is still very efficient and shorter than a full-sized 40-meter dipole.

Vertical Random Wire (Activator's Choice)

This antenna has seen military, commercial, and marine use for decades. It consists of a single 29- to 41-foot wire attached to a fiberglass mast and fed directly with an auto tuner at the base of the mast. The auto tuner should be of a type that has a single wire or balanced output as opposed to a coaxial output.

The auto tuner receives the coaxial signals from the transmitter and tunes the single vertical wire in a few seconds.

The advantage of this configuration is that the auto tuner becomes an integral part of the antenna, so there is minimal loss between the tuner and the antenna, and the coax to the

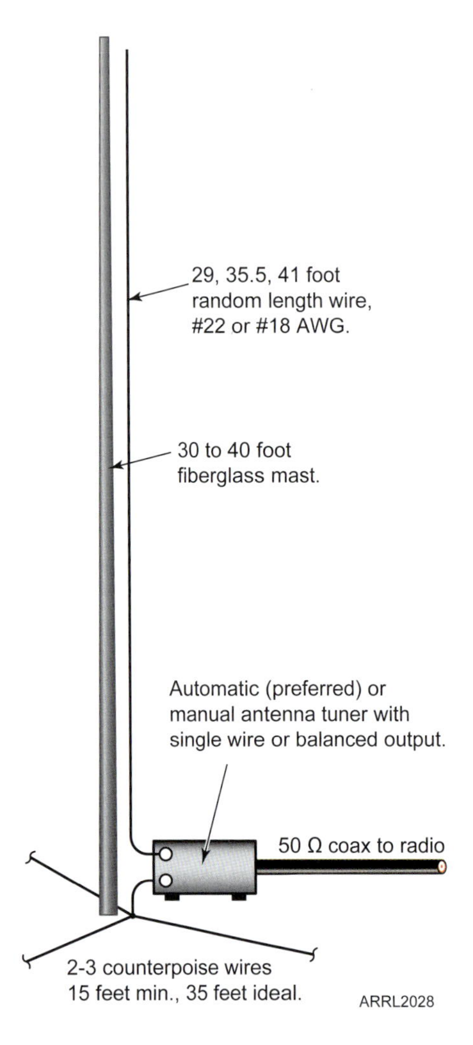

Figure 5, Diagram of the vertical random wire antenna. Note the position of the antenna tuner — directly at the base of the antenna. It cannot be positioned at the radio. Also note the counterpoises (minimum 15 feet). Installing counterpoises dramatically increases this antenna's efficiency.

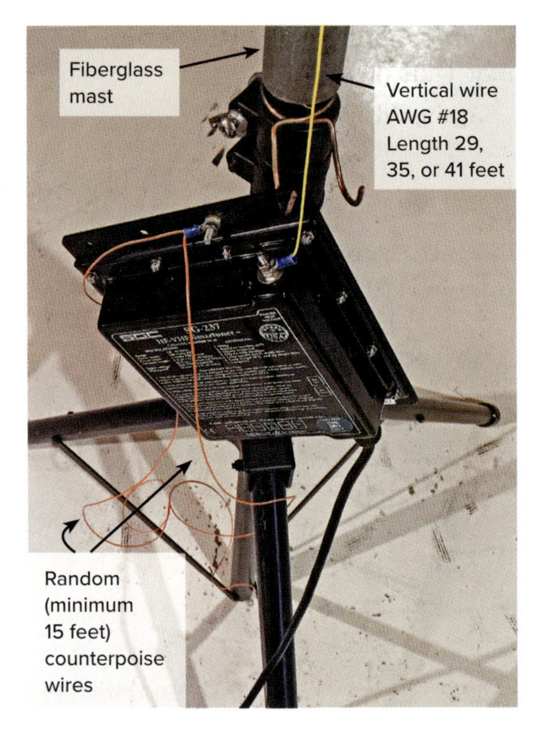

Figure 6, Close up of the vertical random wire antenna's feed point.

station is operating at close to 1:1 SWR, which is ideal.

Radically improved performance is possible if a few random length (minimum 15 feet long) counterpoise wires are laid down on the ground at the base of the mast and connected to the auto tuner's ground connector.

This multiband antenna can operate very well on 40 meters to 6 meters and has adequate performance on 160 and 80 meters. **Figure 5** illustrates the configuration and **Figure 6** shows the vertical random wire antenna's feed point connections.

Nonresonant Doublets or Flat Tops (Hunter's Choice)

Again, old ham radio magazines will discuss this antenna and call it the "universal HF" antenna, fed with a ladder line, or better yet, the "Spencer" dipole, fed with twin coax lines, see **Figure 7**. The Spencer Dipole was described by J.W. Spencer, W4HDX (SK), in the February 1984 issue of *73* magazine. It is very forgiving of lengths of wire, so it can be set up with ease as an inverted-V, upright-L, or as a horizontal antenna if supports are available. For wire lengths, see **Table 1**.

Table 1
Leg Lengths for the Spencer Dipole

Bands Covered (in meters)	*Dipole Leg Length (in feet)*
160 – 10	108
80 – 10	54
40 – 10	27
20 – 10	13.5

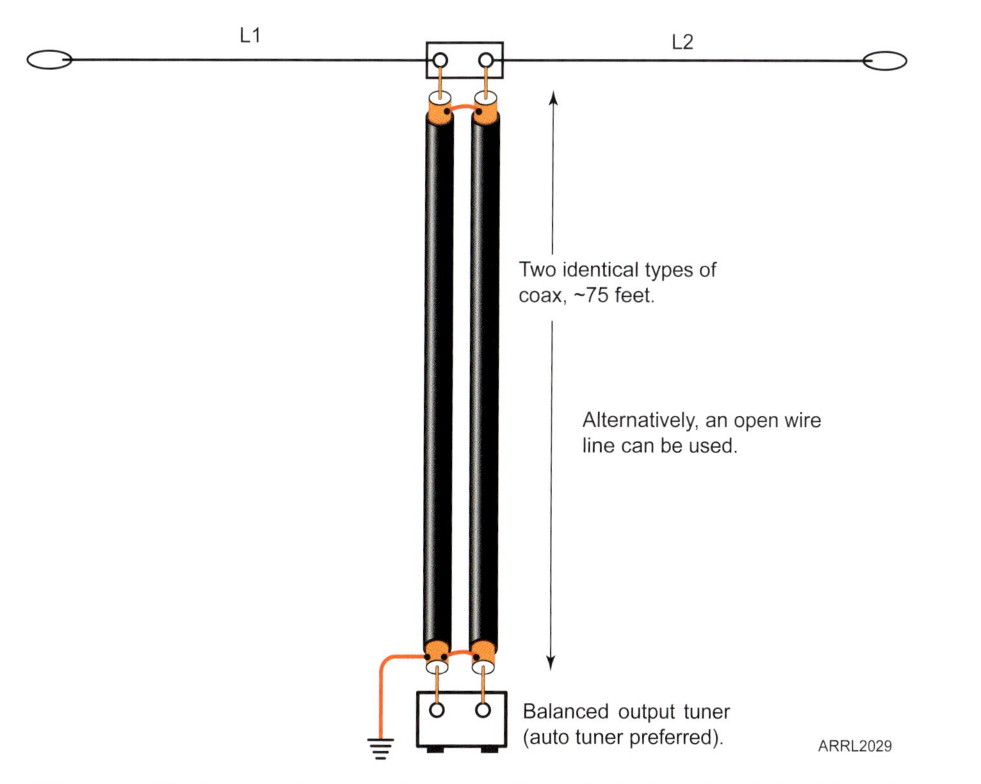

Figure 7, Diagram of the Spencer dipole, a universal multiband dipole. Note that the feed line is two identical lengths of any type of coax. 75 feet covers most bands. If one band does not tune, extend or shorten the feed lines by 3 feet.

Feed the ladder line or twin coax lines with a balanced output auto-tuner and you will have a very efficient multiband antenna. This antenna is a low noise design and my choice for hunting parks.

To use twin coax lines, simply cut two equal lengths of any type of coax (I use RG-8X), and short the braids together at both the antenna and tuner ends. The antenna end braid is connected to nothing and the tuner end braid is connected to earth ground. This twin-coax feed line imitates the function of a ladder line quite well, as long as the twin coaxes are exactly the same length, and electrically longer than ¼ wave at the lowest frequency of operation. See **Figures 8** and **9**.

In the spirit of ham radio, optimizing antennas is the easiest and quickest way to increase your radiated RF power and make activating parks a sure thing at any RF power level. If you have ever worked me at a POTA park, I was likely using 5, 10, or 15 watts SSB

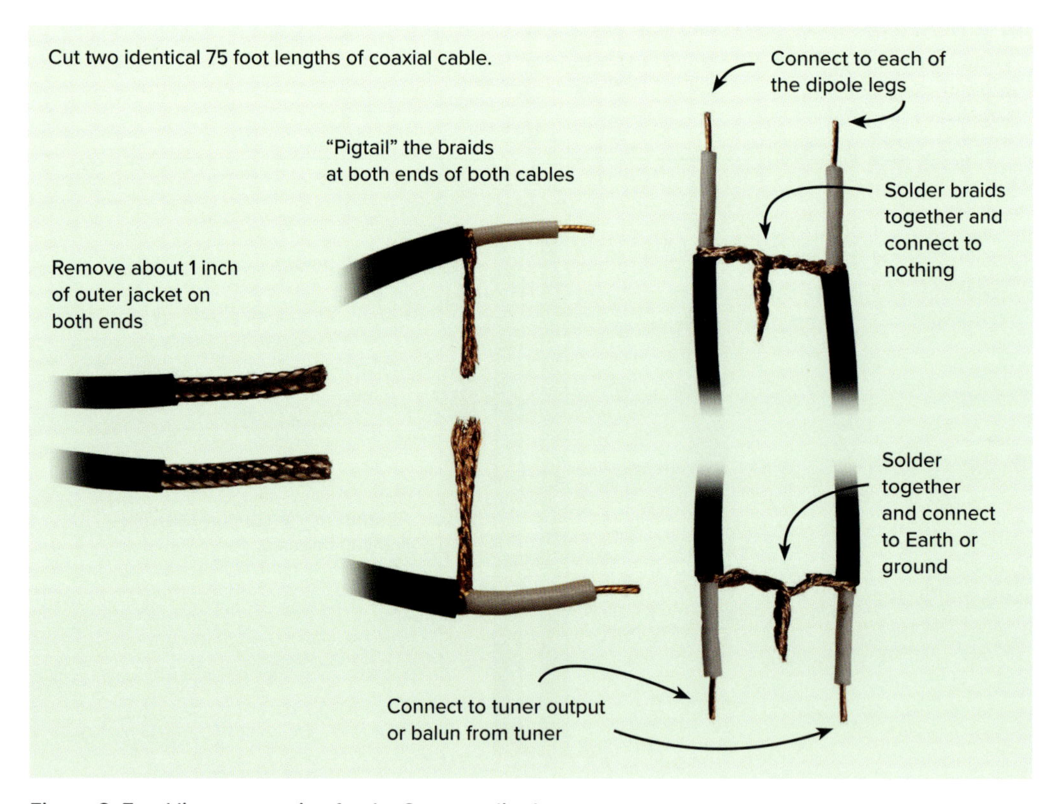

Figure 8, Feed line preparation for the Spencer dipole.

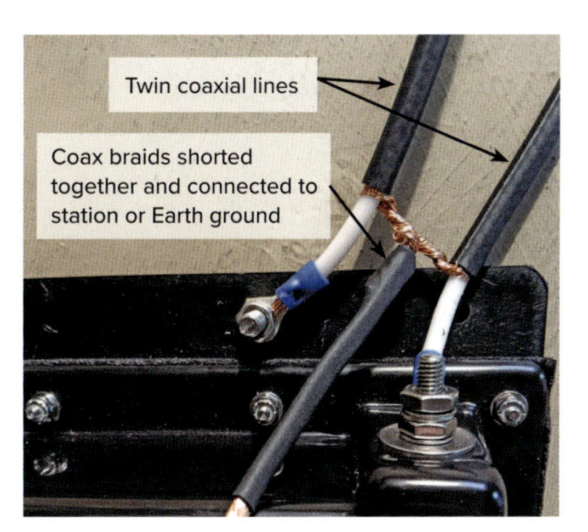

Figure 9, Spencer dipole feed line connections to the tuner. Note that the coax braids are shorted together and connected to the station or Earth ground.

into my upright-L vertical dipole. The only investment I made was to buy a 30-plus foot fiberglass mast that I mount on a DIY trailer hitch mount on my van. My antenna is very efficient, so my 15-watt signal sounds like anyone else's 100-watt station using less efficient portable antennas. Furthermore, since antenna efficiency works both ways, I can hear fainter signals than I would on a less efficient antenna.

Wire antennas have been around a long time because they work. Try wires. Activate with success!

6 Big Benefits from Small Stations

By **Matt Heere, N3NWV**

I've been a POTA activator and an avid motorcycle rider for several years now. Eventually it only made sense to try and combine these two passions, but the Icom IC-7300, 25 amp-hour battery, and 10-meter fiberglass mast simply did not fit on two wheels. My POTA station needed to go on a diet. What I didn't anticipate when I started adapting my station to fit on my bike are all the secondary benefits that come from traveling with as little as possible.

Embrace Serendipity

Because my kit is small and light, I've used it at times and in places I couldn't have had the larger kit. I went fishing with a buddy on Lake Wilhelm in Maurice K. Goddard State Park, K-1381 (in western Pennsylvania). He drove to the lake, and I was already loaded with fishing gear to bring. There's always room for a KX2 and a "random" wire though, so I threw them in with the tackle. The fishing was a bust, but I made 27 contacts. I'm certain the 7300 and the dipole would have worked better, except that they were at home, being too much to bring when the radio wasn't really the point of the day.

The same scenario occurred on a backpacking trip in Cook Forest, K-1345. Camping is another gear-intensive activity, and backpacking seriously constrains both available space and acceptable weight. I managed to find a space for the KX2 and managed seven voice contacts on that outing. Not many, but still seven more than I would have had on the 7300 sitting at home. That outing convinced me to brush up on CW and make my station even smaller.

A QCX mini and 20-meter EFHW have joined the ranks. They make up my new go-to station when there really is no such thing as too small or too light. This rig lives on the motorcycle, right in the tank bag and is ready to go anytime the bike is out on the road (see **Figure 1**.) If my ride takes me past a park, summit, lighthouse, or anywhere else, I have gear and can be ready to make contacts in minutes.

Small Is Approachable

My small setup helps me to be a good ambassador for amateur radio (see **Figure 2**.) One of the greatest benefits of POTA is that it

Figure 1, Bike loaded with the tank bag full of radio gear.

Figure 2, N3NWV operating voice with a minimalist kit based around the Elecraft KX2 transceiver.

gets operators out in front of the public. Hams get some much-needed PR by taking radio to the people. For this to work in our favor though, the public needs to see something relatable. ARRL Field day is a fantastic event, but large multi-operator setups can become a bit of a spectacle with trailers and tents, generators, lots of equipment, wires, and seven people all talking at the same time. It also only happens once a year.

A single operator with a small radio and a thin wire in a tree might arouse the curiosity of a couple out walking their dog, but it probably won't scare them off. More activations than I can count have included a passerby asking "Listenin' to the ballgame?" or "What 'cha up to?" This is my opportunity to tell them about amateur radio. I tell them hams haven't died out and no, the whole world is not on Facebook. If I look like I'm running a nuclear power plant, or hunting UFOs, I either don't get this opportunity, or — worse — confirm existing negative stereotypes of the hobby.

Save Time, Energy, and Money

I save lots of time and energy. It's obvious that the more stuff you pack, the more stuff you must move around. It's less obvious that it also takes time to unpack, move, and repack all that stuff. And time equals contacts. I often run more than 100 contacts per hour at a POTA activation. Assuming I have a fixed amount of time (usually a valid assumption), 20 minutes on either end of my activation is 60+ contacts! My actual setup and teardown time is currently close to five minutes, allowing me to be on the air longer and get more of those call signs into my log. I toss a string

Figure 3, An ultralight kit designed around the QRP Labs QCX Mini CW transceiver.

over a tree limb, haul up my end-fed wire, run 25 feet of coax from the radio to the antenna, turn on the radio and start my operation. Tear-down is the reverse, and this is everything I brought, so I waste no time rummaging through gear looking for what I need.

I also save money. Stuff isn't usually free, and as a rule, bigger stuff comes with bigger price tags. It pays (literally) to make sensible decisions about what capabilities your portable station needs. A POTA activation is not a weak signal hunt. You will likely have a pile-up calling you, so you probably don't need a waterfall, a world-class receiver, or six variable-frequency oscillators. If a 10-watt CQ yields 20 people calling you back, do you really need 100 watts? Radios with simpler feature sets and lower power outputs can be made smaller and less expensive without negatively impacting your log. The same goes for antennas. You will likely make contacts on a small number of bands for a short period of time using 100 watts or less. A do-it-yourself dipole, ¼-wave, or end-fed wire antenna will get the job done for little-to-no money, and wire antennas pack down to nearly nothing.

Finally, the challenge is simply a lot of fun. The goal of shrinking my kit has led me to build my own radios (the QCX mini pictured in **Figures 3** and **4** was a kit), build my own antennas (an EFHW is awesome and simple), and dust off my (admittedly rusty) CW skills. Each of these aspects of amateur radio is entertaining on its own. Taken together — and with a specific purpose in mind — the fun factor just keeps increasing. Making contacts is fun. Making them with tiny gear is even more fun. Making those contacts from a spot that probably no one else is going to is just an over-the-moon good time.

Figure 4, N3NWV's ultralight kit packed.

N3NWV's Gear

Smallest Kit

This is the one that lives in the tank bag on my motorcycle. It's available to me every time I head out, just in case we run across a park for a rest stop. It weighs nothing, is rugged and inexpensive, but confines me to CW on 20 meters.

- QRP Labs QCX Mini transceiver (20-meter band)
- 3 × 18650 Lithium-ion battery pack (stolen from my KX2)
- Putikeeg mini CW key
- #28 AWG wire EFHW for 20 meters with homebrew unun (T50-43 based)
- Arborist line and small weight for wire deployment
- 10 feet of RG-174 coax
- Ear buds
- Paper and pen(s)

Voice and Multiband Kit

This is the kit I grab when I know an activation is planned on a ride, hike, or trip. It allows me to operate voice and work whichever band is hopping that day (see **Figures A** and **B**).

- Elecraft KX2 transceiver, attached paddles, and built-in tuner
- Bioenno 3AHr battery
- Hand mic
- 35-foot random-wire antenna with homebrew 9:1 unun (T140-43 based)

Figure A, The components of N3NWV's Elecraft KX2 kit.

Figure B, The KX2 kit packed.

- Arborist line and small weight for wire deployment
- 25 feet of RG-58 coax
- Paper and pens

Higher Power and eLogging Kit

This is a different definition of a "minimum" kit. This kit is not lightweight, and is only small relative to a trunk full of gear. You'd be hard pressed to have 80 – 6 meters at 100 watts, with a computer for logging and digital modes, in a much smaller footprint though.

- Yaesu FT-891 transceiver and hand mic
- DX Commander Expedition model or Linked 40-meter EFHW antenna. DX Commander offers "no-touch" band hopping and "no tree" deployment. The EFHW packs down smaller but requires the occasional connection/ disconnection of links to change bands and a tree to get it in the air.
- Arborist line and small weight for wire deployment (if using the EFHW)
- Homebrewed 18AHr LiFePO4 battery
- 25 feet of RG-58 coax
- Microsoft Surface (gen 5) tablet/laptop.
- Cheap cordless telephone headset and homebrewed headset adapter with PTT switch
- 3D printed CW paddles

Minimal is Relative

As you can see in the gear list, all these principles have come together to guide my station choices. Even so, there isn't a one-size-fits-all solution for every occasion. Much less so for all operators everywhere. I'm not always on a motorcycle (plenty of rain and winter here in western Pennsylvania) and my CW skills are still fledgling. That's why I have three different setups, but all of them can be considered "minimal" in the scenario for which they are intended.

Take a good hard look at what you really need for your activations. If you can send and receive RF and jot down a call sign, you have enough gear to complete a POTA outing. Consider leaving anything else behind. If it's slowing you down, weighing you down, or — worse yet — keeping you at home, ditch it. Your smile and your log are the only things that should always be big when you activate POTA.

N3NWV on YouTube

To see Matt's antenna builds and woodworking projects, check out his channel at **youtube.com/MattHeere**.

7 Polite Operation

Practices for Being Your Best Self On the Air, With Examples

By **Jeff Zarge, K3JRZ**

A while ago I was doing an after-work activation on 20 and then 40 meters at K-1741, Lums Pond State Park, in my home state of Delaware. I've activated Lums Pond many times. It's a beautiful park with miles of trails circling Delaware's biggest pond. I set up my station from the back of my SUV, as I usually do, and started operating (see **Figures 1**, **2**, and **3**). The bands were hot, and I wound up with good pileups on both 20 and 40 meters. While I was working 40 meters, a Canadian station tried to contact me. I could hear her voice, but I couldn't make out her full call sign due to the other stations calling, but I did hear "Hotel."

I called for "the station ending in 'Hotel.'"

She started to say her call sign when someone tuned up on her, sending a piercing whine that covered any sound of the pileup. I don't know if the person tuning up on her was doing so intentionally — what could possibly be the reason to do so? — but regardless, it was making contact impossible.

I repeated back her suffix, "Romeo, November, Hotel," and asked her to repeat the prefix.

Again, someone tuned up on top of her, coming through louder than before.

She asked, "QSL?" (*Editor's note*: QSL means "Can you acknowledge receipt?" or "I acknowledge receipt." For a list of Q-signals see this book's Appendix.)

I told her I didn't copy and asked the person that was tuning up to stop.

She tried again, and the station tuned up again. Others in the pileup started giving their call signs. As soon as those stations stopped, I hit the push-to-talk (PTT) button and told everyone I was attempting to contact a station that was getting stepped on by someone tuning up on them.

Even so, another station gave their call sign.

Someone — I think the Canadian station — said, "Nooooo!"

Another station jumped in and tried to help, saying, "I think he's trying to work another station. Hold on."

Silence.

I asked for the station ending in, "Romeo, November, Hotel," to give me their call sign prefix again. It took a second request, but she came back, and I heard her.

I made the contact with her, thanked her for her patience, and finished.

Some might ask, "Why didn't you just ignore that person tuning up?" All I could hear was the whine of the station tuning up on her. It was almost enough to make me lose my cool, which can happen in the heat of a pileup. Maybe that op didn't know they were tuning up on a sizeable pileup, but ultimately, whether they knew what they were doing or not doesn't matter. The only ham that any of us can control is ourselves, so it behooves us to do what we can to keep the airwaves respectful. The DX Code of Conduct (DXCoC) is a powerful and popular tool to help every op present their best selves on the air.

Best Practices for Operating

The DX Code of Conduct

During the 2009 VK9GMW Mellish Reef DXpedition Randy Johnson, W6SJ (SK), grew frustrated with the discourteous behavior he heard in the pileups. He and others from the First Class CW Operators' Club and other non-FOC members created the DX Code of Conduct (for operators just starting out or otherwise unfamiliar with CW abbreviations, *DX* means "distance" or "foreign countries"). In 2016 the Yasme Foundation recognized the contribution the code and its website, **dx-code.com**, had made to amateur radio with a grant intended to support the site.

Since then, the DXCoC has spread across the internet in the amateur radio community. It is displayed or mentioned on many amateur radio websites including the Radio Society of Great Britain (**rsgb.org/main/operating/dx-code-of-conduct**), Northern California DX Club (**ncdxc.org/dx-code-of-conduct**) and even on the Parks on the Air website (**docs.pota.app**). Many hams proudly display the logo from the DXCoC's website on their **QRZ.com** profile. If you haven't heard of it by now, here is what it says:

1. I will listen, and listen, and then listen again before calling.

Figure 1, Running *HAMRS* from the back of K3JRZ's SUV.

2. I will only call if I can copy the DX station properly.

3. I will not trust the DX cluster and will be sure of the DX station's call sign before calling.

4. I will not interfere with the DX station nor anyone calling and will never tune up on the DX frequency or in the QSX slot.

5. I will wait for the DX station to end a contact before I call.

6. I will always send my full call sign.

7. I will call and then listen for a reasonable interval. I will not call continuously.

8. I will not transmit when the DX operator calls another call sign, not mine.

9. I will not transmit when the DX operator queries a call sign not like mine.

10. I will not transmit when the DX station requests geographic areas other than mine.

11. When the DX operator calls me, I will not repeat my call sign unless I think he has copied it incorrectly.

12. I will be thankful if and when I do make a contact.

13. I will respect my fellow hams and conduct myself so as to earn their respect.

You might ask, "What does the DXCoC have to do with Parks on the Air?" The DXCoC applies to every aspect of being on the air, every band and every mode — whether talking to a DX station or to another ham across the state or country; on HF, VHF or UHF; on SSB, CW, or a digital mode — the DX Code of Conduct still applies. Most simply, the DXCoC is the ham version of the Golden Rule: treat others the way you would like to be treated. In addition, every POTA activator and hunter tacitly agrees to abide by the DXCoC, because the code is an integral part of operating POTA. It's listed in the documentation right after the rules, and operators are expected to abide by it.

You might ask, "But 'DX' is in its name. Shouldn't it only apply to DX contacts?" DX happens mostly on HF frequencies (at this time I won't address digital voice modes). There really is no difference between a traditional DX contact and an HF contact happening between two stations in the same country, or even in the same state or province. The courtesies we extend to each other should remain the same whether we're talking to a ham around the corner or on the other side of the world.

Being a regular POTA activator, I've had and heard others experience many instances where — let's call them "anxious hams" — disregard the common courtesies of the DXCoC. Being in Delaware, where I do most of my POTA activations, I am sometimes a "DX entity" and anxious POTA hunters from all around the world want to make contact. I don't always have huge pileups, but when I do, most hunters are courteous and polite. However, sometimes when I've called "QRZed" (*Editor's note*: QRZ [or "cue arr zed," as it's usually said], technically means "Who is calling me?" but has come to mean "Next call sign?" when operating a pileup.) I have heard hunter stations ignore many of the best practices promoted in the DXCoC and engage in behaviors like:

- Giving a call sign repeatedly without listening
- Giving out only the suffix of a call sign
- Giving out a call sign when a completely different station has been called
- Giving a call sign when an activator is in the middle of a contact with another station
- Giving a call sign at the end of a contact

K3JRZ's Gear

Radios
- Yaesu FT-DX10 transceiver
- Xiegu G90 transceiver

Case
- Apache 4800 from Harbor Freight (FTDX10)

Antennas
- MFJ 1610T 10M HF Stick
- MFJ 1617T 17M HF Stick
- MFJ 1620T 20M HF Stick
- MFJ 1640T 40M HF Stick
- Shark Antenna S-F6 Ham Stick
- Shark Antenna S-F12 Ham Stick
- Shark Antenna S-F15 Ham Stick
- Buddipole Buddistick Pro
- KM4ACK EFHW
- A few homebrew antennas are being planned

Tri Mag Mount
- MFJ 336S

Batteries
- Bioenno Power 12V, 30Ah LFP Battery (PVC, BLF-1230A)
- Bioenno Power 12V, 9Ah LFP Battery (PVC, BLF-1209A)

Battery Box
- Powerwerx PWRbox Portable Power Box for 12-40Ah LiFePO4 Bioenno batteries

Tripods
- UBeesize 64" phone tripod
- Anozer flexible tripod
- ULANZI MT-33 flexible mini tripod

External Microphones
- DJI Mic Kit
- Movo VXR10-PRO external video microphone

Camera
- iPhone 14 Pro Max

Logging Device
- iPad Pro Gen 2 (Spotting, information, and logging)

Logging App
- *HAMRS*, **hamrs.app**

Ham Stick Bag
- Lixada 55-inch fishing rod bag

Other bags and pouches
- Special operations equipment: **originalsoegear.com** or **soetacticalgear.com**

---

where it is customary to say, "Thank you for the contact and 73!" and the activator is waiting for the station they just worked to respond in kind. Let's give them that chance and wait for the activator to call QRZed again.

Some of the practices listed above are tempting. We're all out there trying to make contacts and build our activator or hunter scores or to achieve an all-time new entity, but when we use the tactics listed above, we're not being respectful of our fellow hams or, more broadly, of the hobby. Following the DXCoC helps everyone share the bands and have a good time.

More Tips for Bringing Your Best to the Air

So we've looked at the DXCoC and some things to avoid while operating. Before we move on to further examples, I'd like to point out a few additional good practices.

- When picking a frequency to start calling CQ, it's customary to listen for stations already using the frequency and then to ask, "Is the frequency in use?"
- If you're using a frequency and someone calls out to find out if anyone is using the frequency, you should politely reply that the frequency is in use.
- When hunting a POTA activator, listen for the activator to say "QRZed." That's how the activator lets you know that he or she is ready to take the next call sign.
- If a net happens to slide onto your frequency or you hear QRM due to changing propagation, be the bigger operator and move frequencies.
- And maybe most importantly, speak and comport yourself politely.

These are just a few common courtesies to observe when on the air. Most hams abide by these simple rules whenever they operate. Sometimes, however, you will run across a hunter or activator who seems not to care about operating well and politely, who, it seems, just wants to snag the contact. When bad behavior occurs, it can be enough to aggravate even the most patient, humble operator. But before you write off the source of QRM as a jerk, it's important to take a moment to reflect on what might be going on. Assuming the other operator is bad probably isn't the best way to support the hobby. Instead, consider the situations your fellow hams might find themselves in, and offer them the benefit of the doubt. Maybe the operator isn't hearing the stations that you are due to some quirk of propagation. Maybe they're having antenna trouble or they're experiencing high levels of atmospheric noise. There are many possible explanations for why people behave rudely. To be the better op, it's best to extend the benefit of the doubt to your fellow hams.

Tales From the Park: Failures and Triumphs

A Benevolent DX Station

About a year after getting onto HF, I was trying to contact an Israeli station that I heard repeatedly on the air during the spring of 2016. I had commented to him on Facebook that I was eager to make contact with him, but every time he was on the air, I just couldn't break through the pileup. The next time he was on, he notified me. I couldn't break through again, and I let him know on social media.

He finished a contact and called my call sign. Other stations started giving their call signs. He said he only wanted to hear from K3JRZ.

I said,

"This is Kilo three Juliet Romeo Zulu."

Again, other stations started to call.

He said he would turn off his radio if people continued to throw out their call signs. Silence. The Israeli station said "K3JRZ" again.

I replied, and he could barely hear me.

Throughout the contact we struggled to pass info back and forth, but in the end, we managed. Making that contact thrilled me. I thanked him for helping me out. It's a contact

I will never forget. It taught me about being the little guy, and about how amazing it can feel to accomplish something you're not sure you can — with some help from a friend. That contact made me want to pay it forward as many times as I can. It's what I try to do on the air.

Persistent Hunter

On a chilly April day in 2023, a European station kept throwing out his call sign in the middle of several contacts I was trying to make. I let him know I was in the middle of a contact and that he should wait.

A few moments later — while I was still in the same contact — the European station gave out his call sign again, and then a few domestic stations threw out theirs. I politely asked them to please not interrupt my contacts with other stations and to please wait until I called QRZed.

Right then, the European station gave his call sign.

I said again, "Please wait until I call QRZed," but the station put out his call sign again before I finished my contact.

Figure 2, Jeff's portable operating position.

I was a bit frustrated by this behavior. I finally called QRZed and had a pileup come back. I did hear the European station, but I was not going to reward him, and I called another station.

During that contact the European station threw out his call sign again, interrupting that contact as well as the one after that.

The way to deal with this sort of unwelcome persistence is to not make the contact, which is how I chose to handle the situation.

I don't know if he just couldn't hear the other stations I was talking to or if he was just anxious to contact me. I can't get inside his head. This is where the first tenet of the DXCoC, "Listen, listen, and listen some more," serves the community well. If we can't hear the station or stations that are hunting the activator, we can listen and wait for the activator to call QRZed.

Net Encounters

No One Owns an Amateur Frequency

Most of us have heard about or experienced being asked to move off a frequency because a net is going to start shortly. While it's perfectly reasonable to shift up or down to accommodate a net, it isn't required by FCC rules. The FCC's website states in Title 47 CFR 97.101(b) (Part 97, Subpart B, section 101(b)) that "No frequency will be assigned for the exclusive use of any station" (**ecfr.gov/current/title-47/chapter-I/subchapter-D/part-97#p-97.101(b)**). I do know many nets on the HF bands that say they'll meet at X frequency +/-, meaning that they'll try to meet at a specific frequency if it's clear and available. They even state that they'll coordinate

Figure 3, Making contacts at Lums Pond State Park, K-1741.

to make sure people find the frequency where they'll have their nets so people can find them if they're not at their "usual" frequency.

I commend those stations. I always call out a minimum of three times to see if the frequency is in use before I start calling CQ POTA. Ninety-nine percent of the time I hear nothing back, but a few times I might get someone stating that there is a net in progress. I apologize, stating that I was checking as I didn't hear anyone from my side, and I gladly move out of their way so as not to further interrupt their net. There will be collisions between nets and other stations from time to time. Sometimes propagation changes and my audio might drift into the frequency a net is using or vice versa. In cases like this, I have been asked to move — both nicely and not nicely. In either case, I apologize and move, as someone else's unbecoming behavior doesn't have to affect me.

On-Air Activity Groups

There are many kind, patient, and generous amateur radio operators out there that are great ambassadors for our hobby. In November of 2022, I had just finished asking three times if the frequency I had chosen was in use. I put my mic down for a second as I was finishing up putting the frequency into my logging app, *HAMRS*, and a station called asking if "anyone was using this frequency right now."

I picked up my mic and said that I was just about to, and if they wanted the frequency that they could have it. The station came back, asked for my call sign, which I gave to him. He stated his name was Budd, that his call sign was W3FF. He was in California and that he was part of "an activity group" that met at a specific time on the frequency and that they gave their signal reports to each other. He asked if I'd like to join them.

I agreed and stated I was about to do a POTA activation and would love to join them. I gave Budd my name, the park I was at, and my state.

Budd said that was great, that I should stick around for a bit to make some contacts and then go up the band five MHz so he could send stations up to me to make more contacts.

I sat back and he called to see if anyone could hear him and wanted to chime in. I heard a few stations starting to come in. They gave signal reports. I heard stations from Virginia, Oklahoma, Mexico, and New Zealand (What? New Zealand? I've never heard a station from New Zealand on HF before!) Budd then told his group of on-air friends that he and I both met at that frequency at the same time and what I was doing. He asked me to put my call out to see who could hear me. I had many of the stations coming back to me stating that they could hear me and gave signal reports. Budd asked if the New Zealand station could hear me. I was gobsmacked! The station heard me and we made contact. I next made contact with a Mexican station and several U.S. stations afterward. After moving up the band five MHz and taking more contacts to complete the activation (I had also spotted myself on the POTA spotting site, **pota.app**), I moved back down to 24.970 and thanked Budd very much for the opportunity to meet him and his friends.

A few months later I ran into Budd and his activity group again. I had unknowingly been using the frequency that they usually meet on and had already made several contacts. Again, he was polite and suggested a compromise to our meeting at the same frequency.

Kindness to Strangers

On the air or off the air, it doesn't take much to be kind to others — to those that have the specific interest of amateur radio or to those you come across in your day-to-day life. Most people were raised to treat others the way they would like to be treated. Amateur radio is about fellowship in a hobby that has thousands of different aspects. We share that same interest of talking to people on the different amateur radio bands, frequencies, modes, and doing different activities related to the hobby. We all studied for examinations to get licensed at different levels wherever we live. It costs nothing to behave well. I believe that being polite and courteous to others is the way to be both off and on the air. That way we can demonstrate to everyone that we are good ambassadors of the POTA program and of amateur radio.

8 36 Parks and 12 States in 21 Days

A POTA Rove

By **Lisa Neuscheler, KC1YL**

For the past five years I have taken a road trip from Tampa, Florida, where I now live, to my hometown of Wilton, Connecticut. Four times I left at 12:30 am, drove for 14 hours, slept in Virginia, then drove another 12 hours the next day. When I arrived in Wilton, I was cranky and exhausted. In 2022, I decided to take an extra day on the road each way, which allowed me to break up the drive into five-to-seven-hour legs. It gave me time to walk around, take pictures, "play radio," and have a snack. In total, I drove almost 4,000 miles — almost double the round trip distance — and activated 36 parks in almost every East Coast state. I wanted scenery, so I routed through the Blue Ridge Mountains of North Carolina and Southern Virginia, rather than up the coast on I-95.

To get on the air, you just need the basics. With that in mind, I packed strategically, and boiled everything down to a couple of cases and a backpack that fit easily in the back of my car.

Planning and Preparation

I planned three or four park activations for each day of the trip, and tried to stay in hotels near state lines, so I could activate the border for the 3905 Century Club (CC) nets. Before embarking, I scoured state park maps to get an idea of where to set up. Many parks were huge, and picking a spot ahead of time saved me a lot of hassle. State park department websites were my main resource for park hours. I tried to time my driving so I'd arrive while parks were open, a requisite for POTA.

First Leg

My first park of the trip was the Hofwyl-Broadfield Plantation State Historic Site, K-3719. Before deciding on the plantation about a month before my trip, I had looked at the POTA map to see how many times it had been activated and to determine if there were any restrictions. Seeing nothing to deter me, I emailed Paul Lourd, W1IP, my ham radio mentor, for suggestions. He had activated the park before and gave me information about where to park and set up and what facilities were there. When I arrived, I stopped at the office, and the ranger was very pleasant and provided a bit of park history, from its time as a rice plantation built by slaves to its passing

KC1YL's Gear

What do you really need for a 12-state, 36-POTA park, almost 4,000-mile, 21-day adventure? This is my standard kit, with a few extras since it was a long trip, and spares might be needed.

Standard Kit

- Yaesu FT-891 transceiver in a waterproof box from Harbor Freight
- 15Ah Bioenno battery
- 9Ah Bioenno battery
- Mobile Bioenno charger from Paradan Radio. (While I drove to the next location, this device charged the battery in the back of the car.)
- Two homebrew end-fed half-wave (EFHW) antennas
- Hamstick antennas (used on either a trunk-lip mount or a tripod) for:
  - 75 meters
  - 40 meters
  - 20 meters (two of these)
  - 17 meters
  - 10 meters
  - Yaesu ATAS-120A antenna (used on a trunk-lip mount)
  - 31-foot Jackite mast in a hitch mount

Figure A, One of my two checklist bag tags. I put the other one on my radio box. They help me make sure I've packed up everything when I've finished an activation.

Essential Supplies

- Backpack
- RG8X coax, two rolls
- Barrel connectors
- Zip ties
- Electrical tape
- Bug spray
- Logbooks
- ARRL Band Chart
- Considerate Operator's Guide
- Pens/pencils
- Caution tape
- Antenna analyzer
- Disposable ponchos
- Small camping table
- Camp stool

Snacks

Any ham radio adventure requires food. I usually pack things that are easy to eat and come in small servings.

- Cheese
- Carrot sticks
- Hummus
- Granola bars
- Crackers
- Water

Sugars and carbs will make you sleepy on the next leg of the drive. One cooler lasts all day. Restock as needed each evening.

to the state in the 1970s. The park was full of mossy old live oaks and had a nature trail as well as a museum in the antebellum plantation house. I set up the hitch mount and an EFHW on my orange Jackite 31-foot mast and made myself comfortable at a picnic table to make some contacts.

Later that day I activated another park, two hours north in South Carolina and after another three hours of driving, I arrived at my destination for the night in Pineville, North Carolina, which is right on the border of South Carolina. That night I checked in to the 3905 CC net from a spot on the state line between North and South Carolina (see **Figure 1**). The 3905 CC nets are open to any radio amateur with privileges on the net frequency, whether calling from inside or outside the US. The 3905 CC nets exist to help amateurs make contacts and exchange QSL cards, usually to achieve Worked All States and with DX entities. The 3905 CC has an award, The Nomad, for being mobile in different states and making a certain number of contacts. So, by setting up on the border I got credit for being in two states at once. As the net was at night, I felt it was safer to operate from a locked car, so I screwed a 40-meter Hamstick antenna tuned to 7.268 MHz into the trunk mount and checked in.

Figure 1, Joining the 3905 Century Club net from the North Carolina-South Carolina state line.

The next day I drove seven hours to Hagerstown, Maryland, hitting two more parks on the way, and that night I checked into the 3905 CC net from the Maryland-Pennsylvania state line.

On my third day of travel, I did a short jaunt to a park in Pennsylvania, then continued to Round Valley Recreation Area, K-7505 in NJ, where I met up with Tobi Massano, AD2CD, a POTA hunter who frequently calls when I activate. The park is huge — 2,350 acres with many parking lots — and contains a large reservoir and a low, rolling mountain, Cushetunk, which “possibly comes from the Lenape, meaning ‘place of hogs,’” according to the park’s website. The mountain rises about 500 feet above the water. I called Tobi when I entered the first parking lot, but she was at another one, so I drove through a few more lots, and when I got to the one she was at, I saw a lady walking toward me waving.

“How did you know it was me?” I asked her.

“You don’t have a front license plate!”

The parking lot was crowded with swimmers heading for the beach area, so we found a quieter place with picnic tables away from all the people. I backed my car into a spot near a shady table and unloaded the hitch mount, radio go-box, EFHW antenna, and 31-foot Jackite mast. I showed Tobi how easy it is to attach the hitch mount: just pop in a hitch pin and tighten some U-bolts. I had designed the hitch mount and had a local metal shop manufacture it.

Setting up the antenna was equally easy,

just attach the end-fed half-wave to the top of the mast and pull out each telescoping section. I used a plastic electric fence post to support the other end of the antenna. The fence posts are great because they're insulators, and if you get the "step-in" kind they've got a little spike on the bottom that you can drive into the ground using your foot.

With this arrangement, it only took about six minutes to set up and get on the air. Tobi was so impressed that after our activation, she ordered a Jackite mast, and when I got back to Florida, I sent her the hitch mount's materials list.

After we met up, I made my way to Wilton, Connecticut, my hometown.

Playing Radio in New England and New York

On my first day in Connecticut, I met up with the Greater Norwalk Area Radio Club for a trip to Sheffield Island Light in Long Island Sound for International Lightship/ Lighthouse Weekend (see **Figure 2**). While Sheffield Island isn't a POTA park, it still made

Figure 2, W1PPL, W1TEG, and KC1YL hanging the antenna from the lighthouse tower.

Figure 3, W1IP, KC1PFH, K1MGD, and KC1YL operating on Sheffield Island.

for an interesting activation. The Norwalk Seaport Association takes care of the lighthouse and graciously ferried Paul LaVorgna, W1PPL; Tom Gunther, W1TEG; Evan Wilhelm, KC1PFH; Martin Diamond, K1MGD; Paul Lourd, and me out to the light without charging us. The lighthouse is a two-story structure of dressed stone. A white, octagonal tower straddles the eastern gable with the date of construction, 1868, in black numerals on the front. Steel railing circles the top of the tower and inside that, there's a 10-paned window for the light. The light isn't used for navigation anymore, but the Seaport Association shines it toward Norwalk for special occasions.

Beside the keeper's house, there was a patio with a party tent shading a picnic table (see **Figure 3**). We set up our transceivers on the table. The association kindly let us climb the stairs up the tower to the top of the lighthouse and hang the antenna from the rail. From up there I could see Long Island and the city of Norwalk, and we made contacts with lighthouses on the New York side of Long Island Sound. What a view!

The next day I headed to Orleans, Massachusetts on Cape Cod, where I would spend a week. On the way, I activated four parks, and during my time on the Cape, I activated three more. On my trip last year, I had visited Nickerson State Park, K-2451, in Brewster, MA, and had met Bernie Meggison, a ham with an expired license. I was determined to visit it again. When I went to the park this year, he recognized me and noticed that my setup had changed and let me know he was studying to get his license reinstated. The Cape Cod Rail Trail State Trail, K-8395, goes through Nickerson, and I found a spot close enough to it that I could get credit for activating both parks. Last year's picnic table was gone, but it was drizzling, so I stayed in the car anyway.

I also activated the Monomoy National Wildlife Refuge, K-0324, where I met Frank Pitzi, N1GDO, another ranger who's a ham. He was very interested in my setup and POTA. Monomoy had sustained severe beach erosion for the last several years, with Cape Cod's famous sandy bluffs crumbling into the sea at the rate of one to three feet per month. They'd had to demolish a 115-year-old dormitory and garage, and to relocate a weather station. Frank helped me find a safe place to operate, by an equipment shed, away from foot traffic and any danger from erosion.

On my return to Connecticut, I hit four more parks and joined the 3905 CC from the Connecticut-Rhode Island border.

Over the next couple days, I spent some

Figure 4, KC1YL and W1PPL operating an "inverted vertical" hanging from the Walkway Over the Hudson.

time in New York, from which my activation on the Walkway Over the Hudson stood out. The Walkway is a pedestrian bridge that spans the Hudson River between the city of Poughkeepsie and the hamlet of Highland. It has spectacular views of the Hudson from both sides of the river. That day, Amtrak trains shimmied beside the river on the Poughkeepsie side, and freight trains ran the Metro North in Highland. Boxcar after boxcar snaked along the Hudson while boats puttered under the bridge. Paul Lourd and Paul LaVorgna joined me for the walk. Out over the river, I tied a tennis ball full of fishing weights to the end of a 20-meter EFHW and lowered it from the edge of the bridge. Hanging 212 feet above the Hudson River, I had an inverted vertical (see **Figure 4**)! It worked great, and the view into Poughkeepsie's downtown and of green-sided Illinois Mountain was spectacular.

From the time I arrived in Connecticut to the time I left, just after Labor Day, I activated 22 parks.

Return to Tampa

My first day on the road home, I put in a long drive to Virginia, where I activated two parks before calling it a night in Roanoke. The next day I continued my journey, stopping in Savannah, Georgia, not activating any parks. I was saving up juice for a couple in Florida, the first of which I hit after three hours on the road on September 8: O'Leno State Park, K-1906. It seemed like an interesting park when I looked it up. The Santa Fe

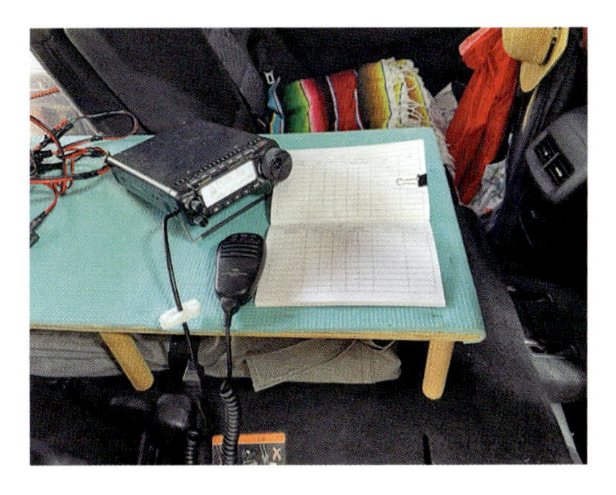

Figure 5, KC1YL's in-car setup.

River descends underground inside the park's borders, only to reemerge three miles later in River Rise Preserve State Park, but that day the weather wasn't cooperating with any plans I had to stretch my legs and take in the scenery. It was a soaker. It rained the entire time I was at O'Leno. Because of the rain I pulled the Hamstick antenna from its case, screwed it into the trunk mount and sealed out the deluge. I screwed the coax into the back of the radio, and I was on the air (see **Figure 5**).

After activating O'Leno, I drove 30 minutes to San Felasco Hammock Preserve State Park, K-3651. It was still raining hard, so I used the Hamstick again. That day, it rained from 6:00 am to 2:30 pm, when I got back home. I knew I wouldn't have another chance to operate in those parks, so a little rain wasn't going to stop me.

If you want to get outside and play radio, either locally or on a trip, get out there and try it. You might even have all the gear already. Local clubs, the POTA Facebook page (**facebook.com/groups/parksontheair**), and other forums can provide ideas, and you might even find folks to meet up with to play radio. Think outside the shack! You'll never know what exciting adventure awaits you. Get outside. Fresh air, beautiful scenery, wildlife, and new friends are waiting for you.

PART 3

Operate Anywhere

Cherry Springs State Park, K-1340 – Potter County, Pennsylvania

Photo by Kody L. Beer, K3KLB

Pikes Peak Trio

POTA, SOTA, and the June ARRL VHF Contest

By **Bob Witte, KØNR**

Normally, I operate ARRL's June VHF contest (**arrl.org/june-vhf**) from my family's mountain cabin west of Colorado's Front Range. It's my big event of the year as a VHF enthusiast. In 2022, however, I decided to combine the contest with two other favorite ham radio activities by operating from Pikes Peak: Summits on the Air (SOTA, **sota.org.uk**) and Parks on the Air (POTA, **parksontheair.com**). Pikes Peak is easy to access, and — at 14,115 feet — it towers over the cities of Denver, Colorado Springs, and Boulder, providing an excellent radio horizon (see **Figure 1**).

Figure 1, Pikes Peak is "America's mountain," towering over the eastern plains of Colorado. This photo was taken from Mount Herman, another popular SOTA summit.

The Plan

Pikes Peak is a difficult place to operate from for a 30-hour contest. The summit closes overnight to the public and the temperature can fall rapidly. Most importantly, the lack of oxygen makes a long activation challenging. To mitigate these difficulties, I planned to limit my operating time to Saturday afternoon. The contest started at noon local time, so my wife Joyce, KØJJW, and I made sure to get there early. We set up on the edge of the parking area, away from where the many summer tourists — especially on the weekend — wander around. I didn't want to bother them, and I really didn't want to be interrupted while making radio contacts.

I always review the rules for each activity I plan to participate in before attempting a multiprogram operation. The VHF contest, SOTA, and POTA are compatible, but there are some differences to consider. For the VHF contest, the exchange was the four-character grid locator for Pikes Peak, DM78, which differed from the lengthy SOTA code, W0C/FR-004. And the POTA reference for Pike National Forest, a different entity entirely, was K-4404. To receive SOTA points, I needed four contacts, and ten for POTA. There was no contact minimum for the VHF contest. Besides working a certain number of contacts, I also needed to operate away from my car to qualify for SOTA, which meant my station had to be backpack portable.

I did not expect to rack up a huge score for the VHF contest. Most contesters operate the entire weekend, and I was limiting myself to just one afternoon. I chose the Single Operator Portable category, intended for portable stations operating with less than 10 watts of power output. My goal was to make at least 50 contacts, primarily using the 6-meter, 2-meter, and 70-centimeter bands. I stuck with FM, SSB, and CW for operating modes.

Before the contest, I heard from Doug Tabor, N6UA, and R.J. Bragg, WY7AA, that they would be on the 1.2 GHz band in Wyoming. So, I took along an Alinco DJ-G7T triband handheld radio and a small Yagi antenna to see if I could work them. I also grabbed a 1.25-meter handheld in case anyone showed up on that band. The entire station proved portable and easily fit into my backpack, enabling a short hike away from my vehicle.

Logging can be a challenge on a summit, so I opted for paper and pencil to keep it

KØNR's Gear

Transceivers

- Icom IC-705 transceiver
- Alinco DJ-G7T triband (2 meters, 70 centimeters, and 23 centimeters) handheld transceiver
- Alinco DJ-296T 1.25-meter handheld transceiver

Antennas

- 6-meter end-fed half-wave antenna
- Arrow II, two-meter, three-element Yagi antenna
- Arrow II, 70-centimeter, five-element Yagi antenna
- 23-centimeter Yagi antenna

Batteries and Accessories

- Bioenno battery
- Camera tripod

simple. Later, I entered this into an electronic log file to submit to the VHF contest, SOTA, and POTA. The requirements for the log files are different for each activity, and it took some editing to adjust the file prior to each submission.

Making Contacts

Joyce decided to skip the contest but did a combined SOTA and POTA activation. She got on the air early and completed her activation before the contest began (see **Figure 2**).

Figure 2, Joyce, KØJJW, operated FM on 2 meters for a combination SOTA/POTA activation.

When the contest started, I fired up the station on 2 meters using FM. Normally, I would use the National Simplex Calling Frequency, 146.52 MHz, but on this activation, I opted to use 146.58 MHz, the North America Adventure Frequency. This frequency is becoming an alternate for simplex use during SOTA especially within range of urban areas, where 146.52 MHz is often busy. I spotted myself on the SOTAwatch website (**sotawatch.sota.org.uk/**) and started calling CQ Contest and CQ SOTA. A camera tripod supported the 2-meter, vertically polarized Yagi antenna (see **Figure 3**). I gathered quite the pileup, and worked the stations calling me quickly, giving out signal reports, my grid locator, and usually "Pikes Peak" as the location. Most

North America Adventure Frequency: 146.58 MHz

This idea started with SOTA W6 Association Manager Rex Vokey, KE6MT, talking with George Zafiropoulos, KJ6VU, and others about designating an alternative frequency to use besides 146.52 MHz. The simplex calling frequency can get overloaded at times and mountaintop use can cause interference with other users on the channel. SOTA enthusiasts in North America kicked this idea around and settled on 146.58 MHz as the North America Adventure Frequency. This idea came out of the SOTA community, but it is intended to be inclusive of other outdoor portable operations, such as Parks on the Air.

A few key points to remember:

- The NAAF is 146.58 MHz.
- This frequency is in addition to, not a replacement for, the National Simplex Calling Frequency 146.52 MHz.
- Local usage will vary depending on specific needs.

Figure 3, The Arrow II three-element Yagi antenna has a threaded hole for easy attachment to a camera tripod.

Typically, when I contacted a station on 144.200 MHz (the National Calling Frequency), we would switch over to 6 meters and 70 centimeters and work on those bands, too. Most of my 6-meter contacts were local, but I did snag a band opening to the Pacific Northwest and worked a few Oregon stations in grid CN94.

I easily contacted Doug and R.J. in Wyoming using SSB on 2 meters and then coordinated for an FM contact on 23 centimeters. After some finagling with the antenna, we barely completed the contact at 151 miles, nearly my best distance on that band.

Ham Radio Ambassador

While I operated, Joyce played the role of ham radio ambassador to any tourist who wandered over. Most people wondered what we were doing and a simple reply of "using ham radio to talk to people all around the state" was sufficient. But we also encountered licensed radio amateurs, many of them new to the hobby and not familiar with SOTA, POTA,

folks were satisfied with that, which meant I only had to give out the SOTA reference, W0C/FR-004, and the POTA reference, K-4404, occasionally. After clearing the pileup, I used the handheld Yagi for 70 centimeters, also with vertical polarization on 446.0 MHz (see **Figure 4**).

I worked quite a few stations on FM then switched over to SSB to see who I could contact. The serious VHF contesters were using this mode, with horizontal polarization on the antennas (see **Figure 5**). I made summit-to-summit contacts with Dan Conway, NØLNT, and Ben Anderson, NØBAA, on Genesse Mountain on 2 meters, about 60 miles away.

Figure 4, Bob, KØNR, operates FM on 70 centimeters using a handheld Yagi antenna.

Figure 5, Bob shielded the microphone to reduce wind noise as he operated SSB on 2 meters with the antenna horizontally polarized.

or VHF contesting. For those folks, Joyce usually spent more time explaining what was going on and how we did it. Because we had received questions during previous SOTA and POTA activations we had created a SOTA/POTA card (see **Figure 6**) that helped.

Results

I was pleased with the results of the activation. The weather was favorable, with temperatures between 40 and 50 °F. We came prepared for snow and blizzard conditions, which can happen any time of the year. The wind picked up occasionally but proved unproblematic.

My goal was at least 50 contacts, and I ended up with 84. Scrubbing this for

Table 1
Contest Results

Band	*Mode*	*Number of Contacts*
6m	FM	–
	SSB	11
2m	FM	41
	SSB	12
1.25m	FM	2
	SSB	–
70cm	FM	9
	SSB	5
23cm	FM	4
	SSB	–
		84 Total
		80 Deduped

duplicates dropped the number to 80 for the VHF contest entry log. (See **Table 1**.) I worked 22 grids, producing a claimed score of 2,130. On VHF doing SOTA and POTA, 80 contacts is excellent. I operated for about three hours total, so that amounted to a little over two minutes per radio contact.

Next Time

I was pleased with the results, but there were also ways I could improve future activations. After reading reports from other contesters, I realized FT8 dominated the contest activity on the 6-meter band. I could have picked up many more contacts and grids using FT8. So, an opportunity for next time is to put together a portable station for FT8 to take advantage of that mode.

I could also have improved my performance on 23 centimeters. When packing, I decided to take along the 23-centimeter gear at the last minute, so I just tossed in the triband handheld (with a maximum output of 1 watt on the band) and a modest Yagi antenna. I could have brought along some of my higher power gear and a better antenna. In particular, SSB capability would have had a huge advantage over FM.

Overall, it was a fun day doing mountaintop radio operating on frequencies above 50 MHz. The increased activity associated with the June VHF Contest combined well with the SOTA and POTA activations. Pikes Peak worked out to be a fantastic location for radio operating. Height above average terrain really does make a big difference on VHF!

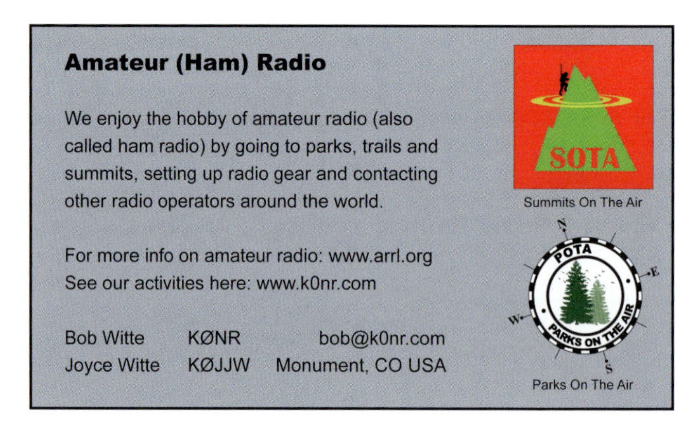

Figure 6, The SOTA/POTA card that we handed out to people who were interested in what we were doing.

10 Urban Park Activations

By **Pete Kobak, KØBAK**

Most sites in the POTA program fit what people generally think of when they imagine a park: an expansive outdoor area away from dense development with grass, trees, and generous parking. However, urban parks often don't conform to that image. Many urban POTA parks are sites of the National Park Service (NPS) that POTA inherited from ARRL's National Parks on the Air (NPOTA) program in 2016. Urban POTA parks can be more challenging than the typical park. These challenges include:

- Lack of designated parking, preventing an activation from a vehicle.
- Needing to walk, use public transportation, or a taxi to reach the park.
- Requiring a pedestrian-portable radio station.
- Crowding and the need to avoid other park users.
- Avoiding radials or guy ropes due to lack of space or because they present tripping hazards.
- Preventing curious park visitors from touching equipment, especially high RF voltages.
- Strict rules against anything in trees or stakes in the ground.
- Tiny operation areas such as single houses with just a stoop or steps outside.
- High and sometimes extreme levels of RF noise.
- Street noise making voice or CW difficult to hear, even with headphones.
- Buildings that disrupt RF, making both transmit and receive signals weaker.
- Highly visible operations that can attract unwanted attention.
- Permission or a formal written permit required beforehand.
- Culturally sensitive sites where radio may be considered disrespectful.
- Security-sensitive sites where law enforcement is suspicious of a radio operation.

Not all items on the list apply to any given urban park, but many items will apply to most sites. Given the challenges compared to the average POTA site, why would you want to activate an urban park? The benefits include:

- Pride in overcoming the many challenges of urban park activations.

• Spots at difficult-to-operate-at parks are rare, which can make your activation popular.
• The constraints of urban park operation hone your station and provide the know-how to activate any park.

Designing a Station for Urban Parks

I gained most of my experience with urban park activations during the NPOTA program in 2016, although I have since activated additional urban parks. POTA, like NPOTA, strictly requires the activation to be on park property (trails being an exception), so my NPOTA experience still applies to POTA. By the end of NPOTA, I had activated all but three sites in New York City, Philadelphia, Baltimore, and Washington, DC.

A Station for Kosciuszko is a Station for All

My initial goal for NPOTA was to activate all units in my Eastern PA ARRL Section (EPA). Investigating the units online, I saw that the Thaddeus Kosciuszko Memorial in Philadelphia was going to be the most difficult. It was a single house in a dense area with narrow streets and was also the nation's smallest NPS unit. Learning that an operator and their entire station must remain on NPS property frustrated me because it seemed to make Kosciuszko impossible.

Even if I had gotten permission to operate from inside the house (which seemed unlikely), the problem remained of using a practical antenna while not damaging the building. I had considered a wire antenna strung up in the dead of night between trees outside the house, but that idea was out because the trees were not NPS-owned. It seemed to me the only possibility would be to operate from the small stoop at one of the doors (see **Figure 1**). The antenna, radio equipment, power, and my considerable frame would all have to fit within an area of about 2 feet by 5 feet. But, if I could solve that problem, I thought I could also fit within and activate every other tiny urban NPS unit in the Northeast as well as activate foot trails. That potential reuse was a powerful motivator. It would not only allow me to achieve my original goal of activating all the EPA Section sites but would provide opportunities for other urban park activations.

Figure 1, Thaddeus Kosciuszko National Monument with magnetic loop antenna and backpack pedestrian-portable station.

Requirements for the Pedestrian Portable Station

Besides squeezing into small operational areas, a key requirement for urban activations was to be able to carry the entire station on city sidewalks for hundreds of yards. Note that this is different than the constraints of an activity like Summits on the Air (SOTA), where an activator must be able to carry their station, sometimes for miles. I aimed to build a station weighing around 40 pounds. That would allow for a 100-watt station and a battery good enough for hours of operation.

The most important part of any station is the antenna, and this is even more true for an urban pedestrian-portable station. The antenna system should be lightweight for carrying, but the lack of available support, or rules against using trees or stakes, prevents the use of simple wire antennas like the popular end-fed half-wave (EFHW). Most urban sites would require a self-supporting antenna system, which meant additional weight for a tripod. I wanted to avoid a loaded vertical antenna because of susceptibility to high urban RF noise, and radials are impractical and hazardous to people at most sites.

These antenna requirements boiled down to two choices among the antennas I was familiar with: a magnetic loop or a rigid loaded dipole. The magnetic loop antenna is naturally low noise, but portable ones have low power handling and can be touchy to tune. A loop is a reasonable choice if one wants to operate with low power, but it's difficult enough to get a signal out between the buildings in a city, so I wanted to stick with a 100-watt station. That left a loaded dipole as the best choice. Efficiency-wise, center-loaded dipole elements would be better, but weight distribution would make the system physically unstable. Therefore, I decided on popular (and inexpensive) products using a thin, several-foot-long, fiberglass-wound loading coil and a tunable whip. The most widely known brand name of this design was Hamstick, but antennas like this were available from several manufacturers. I'll call these "stick antennas" from now on.

While a low dipole would bounce most of its energy upward, getting a little height would help. More importantly, I wanted to keep the high voltage ends of antennas (or radials, if used) out of reach of curious hands in a crowded urban environment. I decided that my existing 10-foot, telescoping tripod would be a good compromise between having minimum height for safety and reasonable physical stability on calm days.

Station Components

Stick antenna dipole mounts with coax connectors were available from a few suppliers. I was careful to get one where one dipole leg was connected to the coax center conductor and the other connected to the coax shield. This mount consisted of four bolts through two square plates, so the dipole could be set up with horizontal or vertical orientation, allowing for additional flexibility in tight locations. An appropriate length of lightweight RG8X cable that easily coils into a backpack completed the antenna system. I didn't find a single rigid carrying case that was both wide and long enough for all the antenna components, so I used the flexible tripod case I already had to carry the tripod and pre-assembled dipole mount and balun. A semi-rigid fishing rod case carried the separate antenna, stick, coil, and whip sections for three bands (12 pieces altogether).

The active part of the system would be the heaviest, so I wanted it to fit into a single medium-sized backpack. eBay offered an aluminum frame system advertised to fit small Yaesu radio models, but the parts of the frame looked adaptable because of the many bolt holes, so I took a chance and ordered one direct from China. While waiting weeks for the shipment, I began specifying the other components. My existing Icom IC-7100 was ideal for this station, having a separate main unit that I didn't have to touch during operation so it could be mounted in the frame and kept in the backpack. I already had a small LDG 100-watt automatic antenna tuner that used Icom's proprietary interface. Therefore, I only needed to purchase a power source.

A lithium (LiFePO4) battery was an obvious choice for its energy-to-weight ratio. Bioenno Power had a good reputation serving the ham community and also emphasized safety, so I looked through their deep-cycle 12-volt offerings for a model that should fit in the frame with my radio. That happened to be a 20 amp-hour battery, which may have been overkill: when I tested it with a 100-watt SSB continuous CQ-and-wait cycle (7 seconds calling, 7 seconds waiting), it lasted more than 5 hours. Along with the battery, I purchased a small voltage booster from a now-defunct vendor, again with an eye toward fitting it in with the other components. Using a booster may also have been overkill for a lithium battery. On the other hand, I had experienced IC-7100 transmission glitches with as much as 12.7 volts supplied, so for the weight and size of the booster, I thought it was reasonable to keep the 7100 fed with 14 volts.

I'm not much of a mechanic, so it was challenging for me to mount everything in the frame. The battery and booster went in the lower half because of weight. The battery rested on the bottom crossbars, with the booster wedged in on one side. I bent steel straps and bolted them between the components and the frame to keep the booster and battery from moving side to side. The battery had a hard plastic case with carrying holes at the top; I drilled into the thin part of the case at the top of the holes to fasten bolts, but I don't generally recommend drilling into battery cases, especially lithium batteries! The radio base and antenna tuner sat on top for easy access to the connectors. Mounting the radio base was straightforward with its existing tapped holes meant for mounting, at least after I found the right metric screw size. To mount the antenna tuner I removed its top cover, drilled holes, bolted the cover to the frame, then attached the body of the device to the mounted cover. In retrospect, it might have been better to tap holes in the cover instead.

The power cabling was simple because only the radio had to be powered. The antenna tuner was powered from the radio connection. I made a short power cable for the 7100 (using their proprietary four-pin connector) with Powerpoles on the other side, which was connected to the output of the voltage booster. I chose an SAE connector between the battery and booster because it matched my battery charger and fit more snugly than Powerpoles. Connecting the SAE plugs was the "on switch" for the system. In the field, a short four-conductor cable was installed between the tuner port on the 7100 radio and the LDG antenna tuner, and a short RF jumper was installed between the radio and the antenna tuner. Although those two jumpers would fit inside the backpack when connected, I was concerned that jostling would bend and weaken the connectors, so I kept them off for transport (see **Figure 2**).

One last space problem had to be solved for the Kosciuszko site. There was no room to sit and still have my feet on the stoop, so I had to operate standing up. I purchased a clever platform meant for a laptop computer that was worn over the shoulders and around the waist to provide a flat surface for the IC-7100 control head and a pad for paper logging. Based on my experience in similar situations, I made a laminated card with my photo ID and ham license to wear hanging from a lanyard. The combination of an official-looking photo ID along with the use of unusual equipment or clothing often kept people from questioning my activity.

Urban Operating

The backpack station worked as designed to activate dozens of urban parks and make hundreds of contacts with park chasers. At most sites I arrived by walking, without planning or permission. However, some exceptional situations required extensive planning or a different type of station. Here are some highlights from my urban park operations that provide examples of special challenges.

Philadelphia

The core of my nearest city, Philadelphia, contains few POTA sites. The difficult Thaddeus Kosciuszko Memorial was my design requirement for the backpack pedestrian portable station. I used a magnetic loop for my first activation, but chasers had difficulty hearing me, which led me to the stick dipole I described earlier. Returning later in the year to activate Kosciuszko again allowed many more hunters to hear me, which validated the decision to switch to a more efficient antenna that could handle 100 watts of RF power.

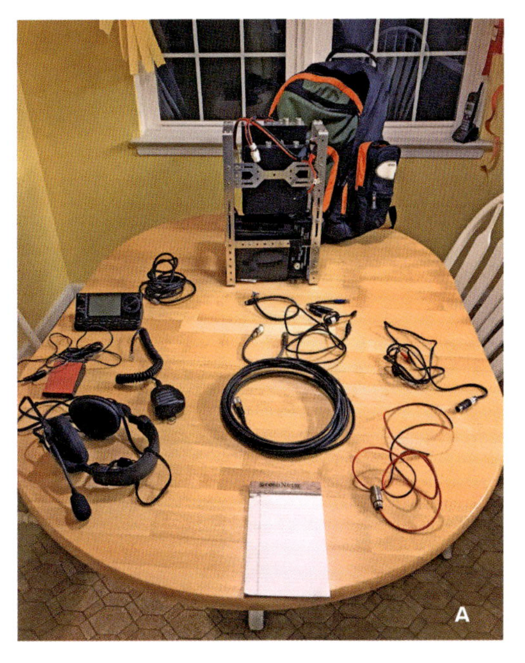

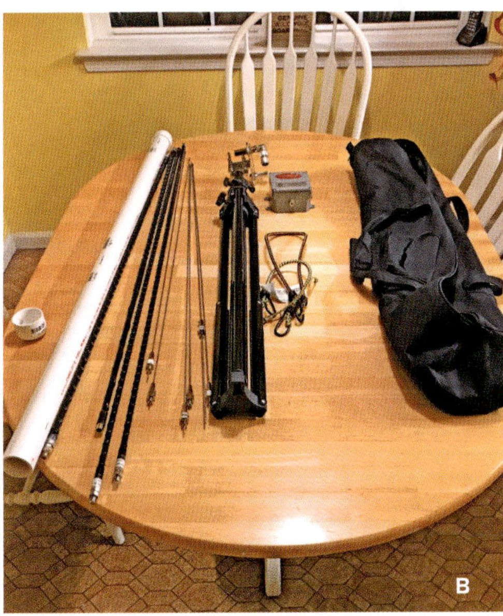

Figure 2, Pedestrian portable station (A) and antenna components (B). Note that in B the antenna case is made of PVC pipe rather than a fishing rod case, as stated in the text.

KØBAK's Gear

Backpack Portable Station

Transceiver and Power

- Icom IC-7100 transceiver
- LDG 100-watt automatic antenna tuner
- 20 Ah Bioenno LiFePO4 battery
- Voltage booster

Antennas

- Dipole stick antennas for three bands (12 pieces), able to mount vertically or horizontally
- RG8X cable
- Tripod
- Tripod case
- Semi-rigid fishing rod case

Backpack Transport System

- Aluminum frame for mounting radios
- Homebrew Icom four-pin-to-Powerpole connector power cable
- Shoulder-mounted laptop platform
- Laminated card with photo ID and ham radio license

Nerd Wagon Add-Ons

- 2-inch-wide mast sections totaling 18 feet
- 500-watt amplifier
- 180 lbs. of sandbags to stabilize tripod legs
- 90 Ah absorbed glass mat (AGM) battery
- Large, garden-type wagon

Figure A, The Nerd Wagon set up for testing before being taken out on activations.

Independence Historical Park and Edgar Allan Poe Historic Site

Independence National Historical Park, K-0738, is the most famous NPS site in Pennsylvania. Independence Mall is three city blocks of mostly open grassy areas, with the main attractions being Independence and the Liberty Bell pavilion on one end. During the busy summer season, I was required to apply for a Special Use Permit. Since I had to have a permit anyway, I decided to build a more powerful station. Still based on a stick dipole, I used a bigger tripod with a taller 2-inch-wide mast to get the dipole up almost 20 feet and sandbag guying. The station included a 500-watt amplifier that I normally used in ham contests when operating vehicle mobile. The station was powered from a deep-cycle

AGM battery and an 80-amp boost regulator. All the equipment was transported in a large metal garden wagon pulled to the site because pedestrian access was required. This station was named "The Nerd Wagon" (see **Figure A**) by park chasers on social media and was described in an article in the June 2017 issue of *QST*.

The Edgar Allan Poe Historic Site, K-0809, was a single house with a lawn, which could also accommodate the Nerd Wagon. Another formal permit was required, for a season when the house was staffed with NPS personnel. A team of three erected the station and operated in the cold rain — which is another thing to plan for if you only have permission for a specific date.

I came back to activate both sites in the cold of late autumn when there were few or no visitors. I could stealthily operate my standard backpack pedestrian-portable station in a quiet corner of Independence Hall and on the stoop of the Poe house.

New York City

Lower Manhattan Trip

I barely planned my first three activations in Manhattan. Originally, I had intended to operate mobile in ARRL's June VHF Contest, but because of technical problems, I had disappointing results, and I was looking for a way to recover my precious weekend time. I had done some research of Federal Hall, K-0773, the Theodore Roosevelt Birthplace National Historic Site, K-0865, and the then-new Stonewall National Monument, K-0977. Due to limited space, I was glad I had a vertical configuration option for the stick antennas at Federal Hall and Stonewall. I also expected tight security at Federal Hall and no security at Stonewall, but both activations upset my expectations.

Federal Hall sat diagonally across the street from the New York Stock Exchange, so naturally I was worried about law enforcement preventing my activation because of the security sensitivity of the area. My major advantage was that I was operating on a Sunday morning when the Exchange was closed, and few visitors were on the streets. I was surprised to set up at the top of the stairs without being challenged. At one point during my operation a uniformed NPS employee passed by on the sidewalk below. I waved to her in a friendly way and she barely acknowledged me. While it was tough getting contacts in the morning and through the winding building-canyons of the Financial District, I was relieved that my operation went smoother than I had anticipated.

The Stonewall Monument was a small area encompassing the tiny Christopher Park across from the famous Stonewall Inn (see **Figure 3**). The Park was the obvious operating location for the Monument. Initially, I didn't think there would be any security issues operating at a small neighborhood park. When I arrived, I was concerned to find at least five NYC police vehicles on the streets bounding the park. I hesitated, but then noticed the park seemed to have normal visitor activities on the warm summer morning. I entered the park and set up on a small brick-paved area. Sitting on the ground rather than a bench so as not to take room from normal park visitors, I made contacts more easily than at Federal Hall. A few minutes into working my pileup, a police officer with heavy body armor and a military-style rifle transited the park in front of me. We glanced at each other, and apparently, he didn't see me as a threat because he didn't even stop to ask what I was doing.

My last stop of the day, at the Teddy Roosevelt Birthplace, should have been the easiest. The classic New York brownstone house was closed for renovations, but the exterior area was unrestricted. I set up with plenty of room and few nearby pedestrians. However, extreme RF noise — exceeding S9 on my radio — as well as high street noise, made activating difficult. I rarely struggle to complete the minimum number of contacts, but at the Teddy Roosevelt Birthplace I almost gave up early. When I was finally done, I was grateful to stop straining my ears.

Governors Island

In the late summer, time was running out to activate Governors Island National Monument, K-0938, because the public ferry from the southern tip of Manhattan only operated in-season. After parking in a large garage on a Sunday when the ferry was free, my first stop was to activate Castle Clinton National Monument, K-0913, in Battery Park. After the Castle activation, I was relieved there was no problem with carrying my station, which includes a suspiciously long case for the antenna system, onto the ferry. Breathtaking views of the Financial District's skyscrapers on that sunny day made the trip itself worthwhile. Only part of the Island was part of the national monument. The core of the monument was Fort Jay, constructed on the island's highest point, so I set up on the grass leading up to the Fort. A beautiful open view to the west over New York Harbor also provided a nice open takeoff area for RF and resulted in a significant pileup all the way to the West Coast (see **Figure 4**).

This happy activation day was ruined when a rogue wind toppled my antenna system, breaking one of the plastic collars on the tripod's telescoping mast. I learned two lessons: don't extend telescoping sections all the way

Figure 3, Operating from Christopher Park, part of the Stonewall National Monument, K-0977.

Figure 4, Governors Island operation at Fort Jay.

so there's some overlap for strength, and be prepared for wind even when the forecast calls for a calm day.

Statue of Liberty

I had been working toward a possible activation of the Statue of Liberty National Monument, K-0963, on Liberty Island, calling and writing to obtain permission. Security control for this site included airport-like screening for visitors boarding the ferry for the island, so it was impossible to bring a significant radio station onto the island without scrutiny. Although I tried to emphasize the small footprint of my one-person operation, NPS officials required me to file a Special Use Permit. With a permit fee of $500 for my relatively short operation, combined with the high costs of a Manhattan operation that included an overnight stay to assure I'd board the first ferry of the morning to meet my NPS contact, this would be my most costly single-site operation.

After a reschedule due to weather, I was excited to finally be the first to activate Liberty on HF for NPOTA on a chilly day in early December (see **Figure 5**). My every move was followed virtually on social media by park chasers hoping to contact the last remaining unactivated NPS site on the East Coast.

A personal NPS escort simplified access to the ferry and the Island. Having learned my lesson about windy New York islands earlier in the year, I brought large water bags for tripod leg ballast that I filled at the NPS ranger station on the Island. I also picked an operation location, specified on the Permit filed months before, that included an informational sign to which I could tie bungee cords for additional wind protection.

Figure 5, Statue of Liberty operation. This was the last un-activated NPOTA unit on the East Coast.

The station setup went smoothly, albeit much later than I had hoped. With every chaser in the NPOTA program needing this site, I experienced the most overwhelming pileup of my ham career. I'm glad my IC-7100 was able to record it. An otherwise exciting and successful activation was marred when I couldn't disassemble one of the antenna sticks to change bands, and then had to stop operating at the strict end time on my permit. Carrying another tool might have allowed me to change bands, but when having to hand-carry all your equipment, it was necessary to balance the chance of needing another tool with the total weight I had to lug around city streets.

Washington, DC

Given security concerns, I had originally assumed I would need permission to activate the many POTA sites in and around the National Mall. Researching the permitting process was daunting since the Mall is host to many events with hundreds or thousands of people. Early in the year I sent in a 12-page permit application, heard nothing back, and though I left messages, no one returned my calls. Later in the year, there were several successful National Mall activations, and they generally reported no issues with law enforcement or park rangers. I decided to plan a visit in October when the weather isn't too cold and there are few park visitors. Two nights at a hotel within walking distance of the Mall allowed me to activate 12 parks by walking between the sites, using my standard backpack pedestrian-portable station.

Figure 6, Late-night Peterson House (part of Ford's Theatre) stoop operation.

One famous DC site not on the Mall, Ford's Theater National Historic Site, K-0814, had not been activated on HF. I was able to contact the specific NPS office covering Ford's Theatre, and after multiple phone calls and redirections, received email permission for a nighttime operation from the exterior stair landing of Petersen House, a part of the Ford's Theatre NPS site. The activation was successful (see **Figure 6**), although on 40 meters one of my transmissions (likely) set off the house's alarm system, which a security guard at the theater reset.

Activate Urban Sites

I hope some of my experiences in building and operating a pedestrian portable station in major cities will encourage you to try urban POTA activations too. While there are certainly challenges, there is also guaranteed adventure in cities that you can't find in the more rural parks. Urban activations like the ones I've described well-deserve the pride you feel when you succeed.

11 The Art of the Self-Sufficient QRP Field Kit

By **Thomas Witherspoon, K4SWL**

One of the things I love about Parks on the Air is that no matter where you travel, you're likely to pass near a POTA entity.

As an avid POTA activator, I've discovered the best way to unlock impromptu POTA opportunities is to have a lightweight and dedicated fully self-contained radio field kit always at the ready. To keep the footprint light and retain ultimate portability and flexibility, I build POTA field kits around low-power QRP radios. [*Editor's note*: low-power, QRP, transceivers generally transmit five watts or less.]

For example, my daughter recently had an hour-long event at a local library, and it was my task to take her there and bring her home afterward. I had planned to simply wait for her, then I realized that a POTA site was only three minutes away. Since I had my mini portable field kit in a side pocket of the car, I was able to drive to the site, deploy (and enjoy) an amazing activation, and finish up with plenty of time to pick up my daughter.

Similarly, I recently took a seven-hour road trip and passed by one of those ubiquitous brown state park road signs. It was near lunchtime, so I pulled into the park, found a picnic table, and set up for a POTA activation while I enjoyed my lunch and some POTA "radio therapy."

Reasons to Go QRP

While running 100 watts through an efficient antenna system will optimize your POTA contacts, I would argue that it's not necessary for programs like POTA unless you live on a remote island.

As a POTA activator, *you are the DX*. Since 2021, I have never used more than five watts to activate a park and have never had an issue.

Going QRP can have an especially positive impact on the size and weight of your POTA field radio kit, because:

- Low-power radios are *smaller* and much *lighter weight* than 100-watt radios.
- Low-power radios use less current, thus can *operate for hours* on lower-capacity, smaller batteries.
- Low-power antennas and accessories also tend to be more *compact* and *lighter weight*.

If you like your big rig but don't want to strain your back or break a sweat hauling it to a picnic table or summit, you might consider going QRP, instead. Low-power field kits can be a fraction of the size and weight of a comparable kit built around a 100-watt radio.

While it's true that POTA makes it easy for mobile or a "trunk" activation with a large transceiver and fixed, mobile antenna, I find setting up a fully portable station in the field so much more rewarding because it introduces more environmental variables and "shakes up" each activation. It moves you out of the parking lot and into the woods, onto the beach, or up to the mountaintop, making each activation a mini adventure. Plus, it builds your emergency communication (EmComm) field skills.

This is why I go QRP for POTA: it's portable, efficient, effective — and fun!

K4SWL's Gear

- QRP Labs QCX-Mini transceiver (20-meter version)
- CW Morse SP4 Paddles
- Sony MDR-EX15AP In-Ear Headphones
- Talentcell Rechargeable 12V 3000mAh Li-Ion Battery Pack
- Rite In The Rain Weatherproof Top Spiral Notebook, 4" × 6"
- GraphGear 1000 Automatic Drafting Pencil (0.9mm)
- Homemade 20-meter End-Fed Half-Wave antenna
- Homemade 10-foot RG-316 BNC-to-BNC feedline
- 20-meter × 2-millimeter Marlow Throw Line
- 8 oz. Weaver Throw Weight
- Red Oxx, The Hound, EDC Pack

Figure A, The contents of K4SWL's pack.

The Self-Sufficient Field Kit

What do I mean by "self-sufficient?" I mean field kits that include *everything* needed to successfully activate a park. These minimal field kits are carefully curated to contain *only* what I need, and little else.

Of course, some "shack-in-a-box" QRP transceivers like the Elecraft KX2 have many of the necessary components built in, such as an ATU, a battery, a speaker, and a microphone. Other low-power transceivers, like the Mountain Topper MTR-4B, are more barebones, requiring external components.

Over the years, I've accumulated a number of small, low-power radios and I love building POTA field kits around them. I can literally take one small pouch containing the radio and essential gear to a picnic table or field, open it, deploy my gear — and just like that — I'm ready to participate in POTA (see **Figure A**).

Start where you are, use what you have, and do what you can, the saying goes, and I recommend doing just that when you begin activating in the field. Then you can take your time to make improvements and adjustments to the kit, by making it even more compact and

Figure 1, K4SWL's QCX-Mini and log before the start of his activation.

Figure 2, Red Oxx's "Hound" side-carry bag (A) with plenty of room for a QRP radio and accessories (B).

more effective to deploy, and perhaps better suited to your own unique style of activating. For example, I've rebuilt antennas with thinner wire, replaced larger keys with smaller ones, employed smaller batteries, and found notepads and pens that better fit the available space in my kit. I also love finding a case, pouch, or pack that has the just-right dimensions and qualities I value for the job — a compact, rugged little pack that, when opened, allows me to keep my components in view and provides easy access when I'm ready to activate.

My Goal: A QCX-Mini Self-Sufficient POTA Kit

Let's face it. The basic components of a self-sufficient kit are the kind of things many hams may already have on hand, namely: a small transceiver with a battery to power it, an input device (key, mic, or computer), an antenna with an antenna support, and a log book and pencil, all tucked into a small pack or pouch that's easy to carry or keep stored in the car.

I recently decided to build a self-sufficient POTA field kit around the QRP Labs QCX-Mini transceiver, which I already own (see **Figure 1**). I chose the QCX-Mini because it's also high-performance: it has handled some of the biggest pileups I've ever managed, and the front end is quite robust. For such a small radio, it sports multiple CW message memories. It's also a mono-band transceiver — and while that may sound, at first, like a limitation, it also means that you can build one dedicated mono-band antenna to accompany the radio and dispense with any sort of matching unit. My QCX-Mini was built for the 20-meter band, which, these days, is my most effective daytime POTA band. And while I own several small, QRP transceivers, the QCX-Mini is one of the most affordable, especially considering that the full kit, aluminum enclosure, and AGC modification can all be purchased for under $80.

Figure 3, The QCX-Mini in use with the CW Morse SP4 paddle.

The Components

Following the transceiver, the first accessory I like to choose is the battery. When sizing up a battery for a 100-watt radio, I check the specifications to make sure the battery has the necessary capacity to meet the current requirements for the radio. The QCX-Mini is an extremely low-current transceiver, needing a miniscule 58 milliamps in receive.

For the transceiver's operation, I chose a Talentcell rechargeable 12-volt, 3,000 milliamp-hour, lithium-ion battery pack. It pairs nicely with this radio, since it's about the same size, has a USB port, and at a modest $24, is particularly affordable. By my calculations, this battery could power me through three POTA activations on just one charge. I'm also a fan of Bioenno LiFePO4 batteries, but the Talentcell pack is more compact, more affordable, and the supplied coaxial plug cables are (at 2.5 millimeters) the exact size the QCX-Mini accepts.

When I choose an antenna for my kit, I first decide if I'll be using an antenna tuner (ATU). If my radio has a built-in tuner – or if I plan to use an external ATU – I'll often turn to a nonresonant antenna like a 9:1 random wire or multiband vertical antenna for frequency agility. In the case of the QCX-Mini, since it's a monoband transceiver, there's no need for an ATU: I built a 20-meter EFHW by cannibalizing a broken antenna from my parts box. The 64:1 matching unit was intact, so I simply cut and trimmed a 63-foot length of 26-gauge wire for resonance on the CW portion of the 20-meter band. I then built a 10-foot-long BNC-to-BNC feed line with RG-316. Fortunately, none of this build was difficult, and it was great to reuse old parts.

Antenna supports are often overlooked, but these can be a crucial component to a truly self-sufficient field kit. If you activate parks in regions that lack trees, for example, you may need a self-supporting antenna or a telescoping mast. In the mountains of North Carolina, where we're blessed with lush hardwood forests, I simply include my "bare-bones" arborist throw line in my kit: 25 meters of 2-millimeter throw line plus an 8-ounce throw weight. I wind up this line on my hand using a "figure eight" around fingers and thumb, then simply secure it with a Velcro cable strap (see **Figure 4**). It's an affordable, no fuss solution that I can deploy easily, and without any kinks or knots!

For a CW key, I wanted something especially compact and precise, so I chose a CW Morse SP4 paddle (see **Figure 3**). While I've a number of portable keys that could have fit the bill, the SP4 paddle is one of the most compact and has an attached cable, which means I can get away with a much smaller radio pouch. The QCX-Mini has no speaker, so I included my favorite pair of $10 Sony earbuds.

For logging, I included a Rite-In-The-Rain spiral notepad and a Graphgear mechanical pencil.

Finally, I chose my trusty Red Oxx "Hound" pack, which houses the entire kit in style (see **Figure 2**). There's even enough room for a *second* QCX-Mini, should I want one. I'm considering another just for 40 meters!

The whole kit weighs in at a grand 3 pounds, 5 ounces – 9 ounces of which is just the throw line and weight (see **Figure 5**). The pack measures 9 inches high by 7 inches wide

K4SWL on YouTube

To see some of Thomas' POTA adventures, check out his YouTube channel at: **youtube.com/@ThomasK4SWL**.

Figure 4, The figure eight winding K4SWL uses for his arborist's line (A) and his wire antennas (B).

by 3 inches deep. I could have easily fit this into an even smaller field pack, but I already owned the Hound, and like it.

To the Field

Of course, before you embark on a long journey to a POTA site, you should deploy your entire station at home to make sure you've included all the necessary components. Having done so, I set my sights on the Zebulon B. Vance Birthplace State Historic Site, K-6856. The Vance site is a special one to me: it's a fairly small historic park with a covered picnic shelter from which one has a pleasant view of hay fields and distant mountains, and it isn't often too busy on weekdays when I'm out and about; for these reasons, I activate it frequently. The staff at this park knows me and encourages me to enjoy the site to its fullest, which I do. There are a number of trees near the shelter that allow for experimentation with a wide variety of wire antennas and verticals alike.

On this particular day, I deployed my 20-meter EFHW in a matter of minutes — three, to be exact. The throw line worked a charm, and the antenna gave me a 1:1 match...score!

Next, I hooked up the QCX-Mini to my fully charged Talentcell battery pack.

I hopped on the air and started calling "CQ POTA" using the QCX-Mini's CW message memory function. After a couple of calls, I was auto-spotted via the Reverse Beacon Network. I logged my first 10 contacts in 10 minutes, thus validating the activation. The band was a little rough, but I stayed on the air about thirty minutes more and worked an additional 16 stations for a total of 26 logged. Talk about fun!

Summary

I derive immense pleasure from tossing one small pouch in my carry-on, in my Everyday Carry (EDC) pack, or in my car, where it stays under the front seat, in a storage pocket or console. That little kit allows me — even on the spur of the moment — to successfully complete a POTA activation with gear that's always primed and ready to deploy (see **Figure 6**). (Part of keeping your car kit ready requires maintaining charged batteries. It's best to swap out your POTA kit batteries on a regular schedule.)

And the truth is, most of us radio enthusiasts already have many (if not all) of the components needed to make a self-sufficient QRP field kit already at hand. Considering that all you need is a transceiver and an input device, an antenna and a means of supporting it, and a logbook (see **Figure 7**) all tucked into a small pack or pouch, such a kit is not out of reach for most hams.

Figure 5, To counterbalance the antenna against tree-branch motion on windy days, K4SWL attaches the throw weight to the line. This adds just enough tension to hold the wire in position.

Figure 6, K4SWL's operating position.

Figure 7, Contacts logged during the activation on paper (A) and on a spot map (B).

So, take a page from my logbook and spend a weekend assembling your own self-sufficient low-power radio kit. Then put it to the test! Following your activation, feel free to make changes and adjustments. As you use your kit, you will soon figure out what upgrades you'd like to make.

With such a field kit on hand, more likely than not, you'll find yourself stopping by POTA sites more often, and, like me, you'll learn to love playing radio in the great outdoors.

12 Ten Thousand Parks and Counting

By **Julien "Clint' Sprott, W9AV**

Discovering POTA

It had been a long time since I visited many of the National parks, and my wife Dani, WC9R, had seen very few of them. So, in 2016 when the National Parks on the Air (NPOTA) program was announced, we decided to put a rig in the car and have our own "DXpedition." Fifty-seven parks and 93 activations later (see **Figure 1**), that whirlwind year came to an end, and the ensuing calm was a bit of a letdown.

The next year I began hearing people on the air making contacts from state parks. After making a few contacts with them, I looked up the POTA website, entered a few details about myself, and discovered that I already had credit toward several awards. That got me started in the POTA program in 2018, almost from its beginning, and at an age where it seemed more prudent to focus on hunting than on activating.

Figure 1, W9AV operating from the Ice Age Trail National Scenic Trail, K-4238, during National Parks on the Air.

Competitive Juice

The first year or two, I was rather casual about hunting, just checking the spots on the POTA website and making a few contacts every day. The old POTA home page listed the top 10 all-time hunters, and I was surprised one day in 2021 to see my call at the bottom of the list. That's when the competitive juices kicked in, and I decided to try and work my way up toward the top of the list.

Being retired and spending most of my time at home surrounded by computers with a modest radio station just steps away, it was easy to check the spots every few hours and make a dozen POTA contacts every day. However, I had reached the point where most of the contacts were with parks I had already worked. So, I wrote a small *BASIC* program that allowed me to type in the park number to see if I needed it.

That got me up to the number two hunter spot within a few months, but it was tedious since most spots were ones I didn't need. I was 1,000 parks behind the top hunter and not gaining on him. The hunt also consumed several hours a day.

With the help of a programmer friend, Jeff Mattox (not a ham, but he should be!), we reverse engineered the POTA spotting page and wrote a PHP program that pulled the spots off the POTA website, checked to see if any were needed, provided an audible alert ("Needed Park!") in my own voice, and sent the spot to my N3FJP Amateur Contact Log. Every hour or so, an alert would come. I would go to the radio, which I left on all day, and click on the spot, which tuned the radio to the frequency. I would often be the first one to work the station. Getting there before the pileup was a big advantage. Seven months later, in April of 2022, I took the lead and continued to outpace the pack, crossing the 10,000-park mark in November of 2022.

Now the alerts come less frequently because I've worked most of the US parks, and many of the alerts are from Europe and Japan — not workable using SSB from Wisconsin. I have started filtering out spots from spotters outside North America, otherwise I would be disturbed all night. I'm back to devoting only about an hour a day to POTA hunting (see **Figure 2**), and the hunting does not involve spinning the dial. Instead, I click on needed spots in response to an alert and wait my turn. I'm often surprised how a signal can be below the noise, but if you just listen for a while, it eventually becomes strong enough to work. If a particular band isn't working, I've noticed activators will often switch to a more favorable one, so I seldom despair of making a contact.

W9AV's Gear

I have two stations, one in Madison, WI, and another 50 miles west of Madison near Muscoda, WI. I use them about equally.

Madison Location

- Icom IC-7600 transceiver
- Icom IC-7300 transceiver
- Two refurbished Lenovo desktop computers networked together for log sharing.
- SPE EXPERT 1.5K-FA solid-state linear amplifier
- G5RV antenna at 30'
- Hexbeam antenna at 30'
- 80-meter EFHW at 20'

Muscoda Location

- Two Icom IC-7300 transceivers
- Icom IC-7100 transceiver
- Three refurbished Lenovo laptop computers networked for log sharing.
- G5RV antenna at 40'
- Hexbeam antenna at 30'
- 80-meter half-wave end-fed inverted L at 30'
- 40-meter half-wave dipole at 40'

I can also run either station remotely from the other for FT8 operation using *AnyDesk* software, making me capable of single-operator four radio (SO4R) operation and occasionally do that with three radios on FT8 and one on CW or SSB.

Figure 2, W9AV at his home station outside Madison, WI.

Advice for Hunters

If you don't have my advantage of being in POTA from the beginning and you lack custom software, it will be hard to become an all-time top hunter. However, there are hundreds of other awards that are easily within reach of anyone, and many have endorsements for different levels. Try to work parks in all fifty states or work all the parks in particular states. Try to be a top hunter in the current year. It is even possible to get a POTA DXCC. The POTA website automatically tracks your progress and provides downloadable PDF certificates for these achievements, all at no cost to you. Focus on whatever mode you prefer. Phone is still the most popular, but digital is advancing rapidly, and there are plenty of CW activators. Some of the top hunters rarely or never do CW or digital, although it helps to be active on all three modes.

In particular, FT8 has become very popular, and with the *JTAlert* program (available from **hamapps.com**), you can set an alert for anyone calling CQ POTA and make a dozen POTA contacts every day. There are now programs that will automate such operation, but POTA has ruled out those contacts. As a hunter you needn't check or submit logs, which I find appealing, and there's little opportunity or reason to cheat since there is no required exchange. I usually give a (true) signal report and my state (Wisconsin). I record the signal report I receive and the park number in my log, but there is no requirement to do so.

Since hunters don't submit logs or other proof of contact, the goal is just to get your call in the activator's log. The activators do all the work, so be sure to thank them for their effort. Generally, hunters and activators are polite, and after a while all the top activators come to know the top hunters. Having an activator recognize your call sign or your voice is a big advantage in breaking a pileup or when conditions are poor.

One advantage of hunting parks is that you don't need fancy equipment. Nearly all my contacts were made with 100 watts and simple wire antennas (see **Figure 3**). A hunter's station is likely better than an activator's, so in general hunters should be able to work anyone they can hear. If there is a large pileup, the average hunter may have to wait for the big stations to make their contact, but most activators continue until they have worked everyone who wants a contact, and that rarely takes more than half an hour. Of course, to work the DX parks and others with poor propagation, high power and a good antenna is a big help. If you miss a contact with a new one, just move on. With the thousands of parks in the system, there will be many more opportunities.

Figure 3, W9AV operating from his Muscoda, WI station.

Contesting and POTA

Most of my first 60 years as a ham were spent chasing DX and contesting, but my station was not adequate to win a major contest, and new DX entities got very rare after about 300 and mainly depended on occasional DXpeditions. The latest minimum in solar activity and the COVID lockdown curtailed DXpeditions and opened the door for the rapid growth of POTA. In many ways it's like a contest going on all day every day, and new parks are being added and activated faster than any hunter can work them. Rarely does a day go by when I don't work a couple of new parks, and so progress is limited only by time and effort.

Each year, POTA sponsors several events: a New Year's Week of casual contacts, Support Your Parks Weekends, which occur periodically throughout the year, and the annual POTA Plaque Event which now takes place the first weekend of June. This last one is for those who like more serious competition. Hunters and activators can earn beautiful plaques in various categories during the event. I usually become an activator for it since I have a nearby park, and it is great fun to be on the other end of a pileup. Seeing things from the viewpoint of an activator makes you a better hunter since you learn how callers break the pileup and get in the log. Most activators start out as hunters, making it a natural way to get started in POTA no matter your previous interest or experience.

13 POTA and Satellites

Good Insurance in a Small Package

by Sean Kutzko, KX9X

There are few things more frustrating in ham radio than setting up at a POTA site and discovering the bands are dead. Every activator has experienced this at least once. These are the times when some form of insurance is helpful to pull off your 10 contacts to get credit for an activation. For me, I found that insurance — as well as an entirely new way to enjoy POTA on its own — in amateur radio satellites.

Satellite operating offers several advantages to the POTA activator. The gear is lightweight and portable. In some cases, you can get by with little more than a dual-band handheld transceiver and a dual-band handheld VHF/UHF Yagi. Satellites are unaffected by fluctuations in propagation, and their schedule and availability can be predicted days or weeks in advance. With a little advance notice to the POTA and satellite community, it's possible to meet your contact requirement fairly quickly. It's also a way for Technician-class licensees to participate in Parks on the Air with their own call sign and make long-haul contacts, not just local contacts on VHF or UHF simplex frequencies.

Satellite Basics

There are around two dozen satellites in use for ham radio contacts. Nearly all satellites transmit on either 2 meters or 70 centimeters and receive on whichever of those bands they're not transmitting. Satellites are mode-specific: some use FM, some use SSB/CW (known as *linear* satellites), and some use data. The FM and data satellites function like an orbiting repeater: only one person can transmit into the satellite at a time. The SSB/CW satellites have between 20 and 60 kHz of bandwidth to tune across, just like an HF band, which allows multiple simultaneous users.

Most satellites are *Low Earth Orbit* (LEO), only a few hundred miles above the earth. Because of their low orbit, they move across the sky rapidly, appearing above the horizon for no more than 15 – 20 minutes at a time. There are several smartphone satellite tracking apps to tell you when and for how long a satellite will be overhead and usable. If you're relying on a smartphone app to track satellites overhead, you'll want to make sure the park you're operating from has good cell service. In parks without service, it's a good idea

KX9X's Gear

- Two Yaesu FT-817 transceivers
- 10 amp-hour battery
- Diplexer
- Boom mic headset
- Dual-band handheld Yagi
- Medium-size camera bag

to load up your app at home and make sure that it's updated with the latest Keplerian elements so you can accurately track the satellite. According to NASA, *Keplerian elements* are, "a set of six independent constants which define an orbit." If they are incorrect, you can be looking for the satellite in the wrong place or at the wrong time. On average, satellites pass overhead six times per day.

The area of coverage of a satellite is called a *footprint*. As the satellite moves, the footprint changes constantly; this means that people you hear at the beginning of a pass may be in a different region than at the end of a pass. Footprint sizes vary, depending on the satellite's orbital height. Generally speaking, FM satellites have a coverage area of around 5,000 kilometers across, and the SSB/CW satellites have roughly an 8,000-kilometer footprint. The lower a satellite pass is above your local horizon, the greater the distance you can cover across the footprint, as you can work from one edge of the footprint to the other.

Because satellite passes are short, there's little casual conversation. Simple exchanges lasting no more than a few seconds are standard practice, which lends itself well to a POTA activation!

It's important to verify you're making it into a satellite. Being able to monitor your own transmitted signal through the satellite in real time is known as *full-duplex* operating.

Equipment Needed

Getting started on satellite operating is fairly straightforward. FM satellites offer the easiest path to getting on the air. Any dual-band handheld will get you started. You will need an antenna that offers good gain on both 2 meters and 70 centimeters. Most portable satellite operators use one of two commercial antennas: The Arrow antenna, which is three elements on 2 meters and seven elements on 70 centimeters on a common boom. The Elk antenna is a five-element log-periodic design. Both are lightweight and easy for most hams to point at the satellite as it moves across the sky. Many hams have started on satellites with a hand-held dual-band Yagi and a low-cost handheld. While not optimal, it will get you on the air for little investment.

SSB satellites are more complex and require more gear. You will need a combination of two all-mode radios that can transmit and receive on 2 meters and 70 centimeters at the same time. A common portable satellite station is two Yaesu FT-817s/818s or Icom IC-705s and a LiFePO4 battery. In reality, any two radios that can get you on 2-meter and 70-centimeter SSB at the same time will work (see **Figure 1**).

Both SSB and FM operation will require a diplexer installed on the antenna feed line between the antenna and the radio. With your transmitter, receiver, and antennas in such close proximity, receiver front end overload, or *desense*, can occur when you transmit. The diplexer isolates your transmitted signal from your receiver, eliminating desense.

Figure 1, Two dual-band handhelds and a modified Arrow dual-band Yagi were all I needed to make contacts from Theodore Roosevelt National Park in North Dakota (K-0065), including with Burt, FG8OJ, on the island of Guadeloupe in the Caribbean Sea. (Nancy Livingston, N9NCY, photo)

Satellite POTA Operation — Things to Consider

Find an operating site with a clear view of the horizon. Trees and buildings can interfere with your signal path, especially on 70 centimeters. Use high-quality coax, as feedline losses on VHF/UHF are substantial, even in short lengths. I recommend a minimum of LMR240. Finally, many satellite operators exchange Maidenhead grid squares on satellite, so know what grid square your park is in. The ARRL rounds up useful resources on locating your grid square at **arrl.org/grid-squares**.

Satellites in Action at Park Units

With lightweight radios and antennas, it's easy to throw satellite gear in the car for your upcoming POTA activation. In June 2022, I did a satellite-only activation of Warren Dunes State Park in Michigan, K-1552. The park is situated right on the eastern shore of Lake Michigan, giving a beautiful view of the western horizon from south to north. With no obstacles, I was able to hear satellites in that hemisphere the second they popped up over the horizon. I set up 20 feet from the water and made 27 contacts that leisurely afternoon over five satellite passes, in between

KX9X on YouTube

Search for "KX9X Satellite" on YouTube to find many videos featuring Sean explaining how to work satellites.

Figure 2, A beachfront location means no obstacles on the horizon! I made 27 satellite contacts in a casual afternoon of operating from Warren Dunes State Park, K-1552, in Michigan in between swimming in Lake Michigan and napping in the sun. (Sean Kutzko, KX9X, photo)

Figure 3, Nancy, N9NCY, used two Yaesu 817s in a camera bag to work satellite passes from Logan Pass in Glacier National Park, K-0028, in Montana in July 2022. (Sean Kutzko, KX9X, photo)

swimming, eating, and working on my tan. Contacts ranged from California to Maryland and Florida, with a couple DX contacts in Mexico thrown in for good measure (see **Figure 2**).

In July 2022, my partner Nancy, N9NCY, and I travelled to Montana to enjoy several days in Glacier National Park, K-0028. While the scenery is majestic, finding a clear view of the horizon among the 10,000-foot mountain peaks in Glacier can test even the most experienced satellite operator. I wound up having a successful, contact-laden pass. It is under these circumstances that knowing the trajectory of the satellite can help you choose your operating site. You don't need a good view of the horizon in every direction, just a good view in a specific satellite's trajectory over your local sky. Through strategic use of data from a good satellite tracking app, we were able to find acceptable operating sites in the park. For example, we made many satellite contacts from one of Glacier's most notable features, Logan Pass, elevation 6,647 feet. This included Nancy's first official POTA activation, making 10 contacts from the Logan Pass parking lot through a linear satellite whose path above our horizon that day happened to be split between two mountain peaks for most of the pass (see **Figure 3**). And with only 15 minutes per pass, there was plenty of time to enjoy hiking and seeing what the park had to offer in between passes.

Satellite operating is a great addition to any POTA activation. The gear is lightweight and it can provide insurance for your activation in case you have poor propagation. And because a lot of satellite gear is handheld, you don't have to worry about issues with raising antennas in more sensitive park locations. I encourage you to spice up your POTA activity with satellite operation. Give it a try.

14 365 Days in the Life of a POTA Activator

Chasing the Bailey-Sprott Challenge

By **Kerri Wright, KB3WAV**

Starting Out

Toward the end of 2021, I was contemplating what POTA-related challenges I would set for myself in the coming year. Over the past six years I have activated parks almost daily, but I'd never tried to stick to an extended activating schedule. For 2022, I decided to try for a full year of daily activations. The adventure began as a personal goal. At the time, I was unaware of the Bailey-Sprott Park-A-Day Challenge (**khk.net/wordpress/2023/02/17/the-bailey-sprott-challenge/**). The challenge was named for the two hunters, Ken Bailey, N5HA, and Clint Sprott, W9AV, who hunted at least one park each day during 2021. The challenge didn't initially include activators, but as word of the achievement spread, I, and several other activators, decided to try for it.

Although still working a full-time job, I figured that daily activations would not be too difficult, since I lived in a "target-rich" area. In years past, I had benefitted from activating the many parks in Maryland and its surroundings. The closest park to my home was Morgan Run Natural Resource Management Area, K-6393, about 12 minutes away, which I sometimes used for Late Shift activations. There were several parks near my work as well, which allowed me to do after-work activations when other commitments prevented

Figure 1, KB3WAV's typical setup for activating.

KB3WAV's Gear

Typical Multi-Park Rove Equipment

- Icom IC-7300 transceiver
- Eagle One vertical antenna for 80 – 6 meters
- External LDG tuner and 4:1 balun
- 15 or 20 amp-hour Bioenno LiFePO4 battery

Additional Equipment

- Laptop for FT8
- Palm Pico paddle or Begali key for CW
- Yaesu VX-8DR handheld transceivers (for FM contacts)

a longer rove. If I had more time in the afternoon, I would head west to several parks outside of Frederick, Maryland that have good elevation and are close to one another, making it easier to complete a multipark activation even after a full day of work.

I usually activated parks from my vehicle, using the Icom IC-7300, an Eagle One vertical antenna which worked on 6 – 80 meters with an external LDG tuner and 4:1 balun. A 15 or 20 amp-hour Bioenno battery powered everything (see **Figure 1**). All the equipment usually stayed in the vehicle, so I was prepared to activate at any time. This was a great setup for multi-park roves, which I prefered. However, even with everything living in my car, I found it was easy to forget a vital piece of equipment when I transitioned between vehicles. After a long day of work once, I rolled into my only planned activation for the day. A thunderstorm was moving in, and I needed

Figure 2, The grocery-bag antenna-tie setup.

to finish the activation before it arrived. I quickly pulled into the park and went to get the antenna out of the trunk and into the trailer hitch and oops! The trailer hitch was missing. With no time to run home and grab the hitch, due to the storm, I found an old plastic grocery bag and ended up tying the antenna into the trunk of the car (see **Figure 2**). I wedged some heavy items around the base of the antenna to help keep it in place and hoped the wind wouldn't pick up with the storm. It wasn't ideal, but it allowed me to complete the activation I needed for the day. Looking back, the weather was a constant consideration. Snowstorms, ice storms, thunderstorms, high winds, or even just rain would require changes to my plans or advanced preparation. More than once I made activations work despite less-than-optimal conditions.

Chasing New Parks

Most activators get excited about the idea of adding a new park to their list of activations, and I am no exception. But racking up new parks requires travel — at least some. The first major trip I took in 2022 was a quick one out to Illinois for a grandson's birthday. My husband Ray, KC3RW, and I found several unique parks along the way, and on our way back to Maryland, I convinced Ray to take a short detour into Wisconsin to explore several more new parks. Ray and my daughters now tease me about planning road trips around new parks rather than around taking the most direct route. On longer trips like the one to Illinois, I bring along all my usual gear but supplement it with an extra HF radio, more coax cables, an antenna analyzer, and equipment to make repairs, even if we're using our regular activation vehicle — which sometimes we aren't (more on that in a bit).

HamCation in Florida provided another opportunity to add parks to our growing list and allowed us to meet some old friends and make some new ones in the POTA tent. While I planned our route for daily activations, I also looked for parks with few activations. I enjoy putting a lightly activated park on the air for the hunters. It helps them add unusual parks to their lists as well. Shocco Creek State Game Land, K-6951, and Sandy Creek State Game Land, K-6948, only had a couple prior activations, probably because they are off the beaten path. I also had the chance to do a multipark rove during HamCation with Bill Brown, K4NYM, and James Mon, KE8PZN, both of whom were going for the Bailey-Sprott challenge as activators too. As HamCation was ending, I convinced Ray we needed to add more states to our activated list, which also meant more unique park activations. Our trip home from HamCation to Maryland took the scenic route through the panhandle of Florida followed by parks in Alabama, Mississippi, Louisiana, Arkansas, Missouri, Tennessee, Kentucky, and, finally, West Virginia. I'm pretty sure that was the most direct route for a POTA activator in search of new states!

In October of 2022, I traveled to the southwest corner of Montana, which just happened to be in driving distance of several parks in Idaho and Wyoming as well. I wasn't using my typical setup since I flew out to Montana and wound up renting an SUV. I brought a SOTABEAMS Tactical Mini mast with a 33-foot wire that I wound around the mast with an LDG 4:1 balun and a couple of 33-foot radials. The battery was a 12 amp-hour Bioenno battery to meet airline restrictions, along with a dc inverter to charge the battery as needed for the activations on the multistate and -park rove. However, when I arrived, Montana was getting its first snowstorm of the year and the first park was about two hours away in good weather. Even though I'd activated a Maryland park as a precaution the night before (in case I got hung up while traveling or wasn't able to activate a park the following day for any reason), I decided to brave the elements and head out on the planned rove. Since I didn't have a trailer hitch to use on the rental, I adapted my antenna deployment with a bungee cord and a fence post at Henrys Lake Airstrip State Recreation Area, K-7816. At Henrys Lake State Park, K-2237, a bungee cord, the vehicle mirror, and a pile of snow held the mast upright in the wind (see **Figure 3**). Henrys Lake was beautiful, with

Figure 3, Activating setup at Henrys Lake State Park, K-2237.

the snow on the mountains and the wind whipping up waves on the lake. Thankfully, my rented SUV handled the eight inches of snow relatively well. I finished the three-state rove with Yellowstone National Park (WY), K-0070, and Custer Gallatin National Forest (MT), K-4501, in one day. Over the next couple of days before returning home, I activated several more Montana parks.

More Than Contacts

POTA has allowed me to build relationships with those I've contacted over the air, something I really value. Sometimes I even get to meet them in person, at a park or a hamfest. The hunters are incredible; they play vital roles for activators, beyond the obvious requirement of making a QSO. When planning a rove into an unfamiliar area, with unknown cell reception, I often gave several regular hunters a heads up about my planned route. They would help spot me when I could not, gave me info about available park-to-parks, and it was their help that made for many successful activations. Also, hunters willing to chase the "repeat offenders" made activating multiple parks on a consistent basis much easier, as I continued the quest for 365 daily POTA activations. In addition, I activated many of my local parks repeatedly throughout the year and completed several Eagle's Nest awards for activating a park 100 times.

Finally, family support made this challenge possible for me to complete. Ray really encouraged me in this effort, whether it was going along on the longer roves with me (see **Figure 4**), helping to fix equipment, or just being ok with coming home to an empty house because I was out at a park. Our normal family plans were tied to ensuring I could get in at least one activation each day. Vacations with our daughters and their families usually meant camping at a park or going to a location where I knew I had parks to activate easily. Achieving the Bailey-Sprott Challenge as an activator required a team effort with my family, friends, and fellow POTA hunters and activators all playing a role. Congratulations to my fellow recipients of the first ever Bailey-Sprott award, both activators and hunters, for completing a challenging year!

Figure 4, Ray Wright, K3CRW, activating Tar Bay Wildlife Management Area, K-7751.

Afterword

By **Sean Kutzko, KX9X**

Operating amateur radio from the great outdoors is nothing new. Hams have been taking their gear into the field for a century, and ARRL formally launched its annual Field Day event in 1933. As technology improved and radios got smaller, it became easier to grab your radio for an enjoyable afternoon or to bring it on a camping trip.

Living in apartments for over 10 years, I used portable operating to stay active on the air. As an ARRL staffer, when I realized 2016 was going to be the centennial of the National Park Service, I knew we could use ham radio to help promote their event. After writing the rules and a formal proposal, and getting approval from the ARRL Board of Directors, I embarked on what would be almost two years of administering one of the largest ham radio events ever envisioned, and one of the most popular: National Parks on the Air (NPOTA). Over one million radio contacts were made by hams from 470 National Park Service units in 2016.

Parks on the Air had been around since 2010, but it was still in its infancy. After NPOTA exploded and the event ended in 2016, hams didn't want the party to end. NPOTA was never designed or expected to be more than a single year, celebrating the National Park Service. It was early in 2017 that I received an email from Jason Johnston, W3AAX, who asked about extending NPOTA. When he learned it wasn't going to be extended past 2016, he started to organize Parks on the Air, drawing inspiration from NPOTA and other "On the Air" programs that focused on outdoor, portable operating. As all can see, it's exploded into a worldwide phenomenon, with a great community of operators sharing technique and technical know-how with new participants every day. I smile whenever I activate a park these days.

It's been great to see Parks on the Air thrive. Congratulations to all who helped make it happen, and to all the operators who participate and help others get started. I hope to work you from a park soon!

Appendix A

Q Signals

These Q signals most often need to be expressed with brevity and clarity in amateur work. (Q abbreviations take the form of questions only when each is sent followed by a question mark.)

QRA What is the name of your station? The name of your station is ______.

QRG Will you tell me my exact frequency (or that of _____)? Your exact frequency (or that of _____) is _____ kHz.

QRH Does my frequency vary? Your frequency varies.

QRI How is the tone of my transmission?
The tone of your transmission is _____ (1. Good; 2. Variable; 3. Bad).

QRJ Are you receiving me badly? I cannot receive you. Your signals are too weak.

QRK What is the intelligibility of my signals (or those of _____)? The intelligibility of your signals (or those of _____) is _____ (1. Bad; 2. Poor; 3. Fair; 4. Good; 5. Excellent).

QRL Are you busy? I am busy (or I am busy with _____). Please do not interfere.

QRM Is my transmission being interfered with? Your transmission is being interfered with (1. Nil; 2. Slightly; 3. Moderately; 4. Severely; 5. Extremely.)

QRN Are you troubled by static? I am troubled by static _____ (1-5 as under QRM).

QRO Shall I increase power? Increase power.

QRP Shall I decrease power? Decrease power.

QRQ Shall I send faster? Send faster (_____ WPM).

QRS Shall I send more slowly? Send more slowly (_____ WPM).

QRT Shall I stop sending? Stop sending.

QRU Have you anything for me? I have nothing for you.

QRV Are you ready? I am ready.

QRW Shall I inform _____ that you are calling on _____ kHz? Please inform _____ that I am calling on _____ kHz.

QRX When will you call me again? I will call you again at _____ hours (on _____ kHz).

QRY What is my turn? Your turn is numbered _____

QRZ Who is calling me? You are being called by _____ (on _____ kHz).

QSA What is the strength of my signals (or those of ____)? The strength of your signals (or those of _____) is _____ (1. Scarcely perceptible; 2. Weak; 3. Fairly good; 4. Good; 5. Very good).

QSB Are my signals fading? Your signals are fading.

QSD Is my keying defective? Your keying is defective.

QSG Shall I send _____ messages at a time? Send _____ messages at a time.

QSK Can you hear me between your signals and if so can I break in on your transmission? I can hear you between my signals; break in on my transmission.

QSL Can you acknowledge receipt? I am acknowledging receipt.

QSM Shall I repeat the last message which I sent you, or some previous message? Repeat the last message which you sent me [or message(s) number(s) _____].

QSN Did you hear me (or _____) on _____ kHz? I did hear you (or _____) on _____ kHz.

QSO Can you communicate with _____ direct or by relay? I can communicate with _____ direct (or by relay through _____).

QSP Will you relay to _____? I will relay to _____

QST General call preceding a message addressed to all amateurs and ARRL members. This is in effect "CQ ARRL."

QSU Shall I send or reply on this frequency (or on _____ kHz)? Send or reply on this frequency (or _____ kHz).

QSV Shall I send a series of Vs on this frequency (or on _____ kHz)? Send a series of Vs on this frequency (or on _____ kHz).

QSW Will you send on this frequency (or on _____ kHz)? I am going to send on this frequency (or on ____ kHz).

QSX Will you listen to _____ on _____ kHz? I am listening to _____ on _____ kHz.

QSY Shall I change to transmission on another frequency? Change to transmission on another frequency (or on _____ kHz).

QSZ Shall I send each word or group more than once? Send each word or group twice (or _____ times).

QTA Shall I cancel message number _____? Cancel message number _____

QTB Do you agree with my counting of words? I do not agree with your counting of words. I will repeat the first letter or digit of each word or group.

QTC How many messages have you to send? I have ______ messages for you (or for _____).

QTH What is your location? My location is _____

QTR What is the correct time? The correct time is _____

QTV Shall I stand guard for you? Stand guard for me.

QTX Will you keep your station open for further communication with me? Keep your station open for me.

QUA Have you news of _____? I have news of _____.

ARRL QN Signals

QNA*	Answer in prearranged order.
QNB	Act as relay between _____ and _____.
QNC	All net stations copy. I have a message for all net stations.
QND*	Net is Directed (Controlled by net control station.)
QNE*	Entire net stand by.
QNF	Net is Free (not controlled).
QNG	Take over as net control station
QNH	Your net frequency is High.
QNI	Net stations report in. I am reporting into the net. (Follow with a list of traffic or QRU.)
QNJ	Can you copy me?
QNK*	Transmit messages for _____ to _____.
QNL	Your net frequency is Low.
QNM*	You are QRMing the net. Stand by.
QNN	Net control station is _____. What station has net control?
QNO	Station is leaving the net.
QNP	Unable to copy you. Unable to copy _____.
QNQ*	Move frequency to _____ and wait for _____ to finish handling traffic. Then send him traffic for _____.
QNR*	Answer _____ and Receive traffic.
QNS	Following Stations are in the net.* (follow with list.) Request list of stations in the net.
QNT	I request permission to leave the net for _____ minutes.
QNU*	The net has traffic for *you*. Stand by.
QNV*	Establish contact with _____ on this frequency. If successful, move to _____ and send him traffic for _____.
QNW	How do I route messages for _____?
QNX	You are excused from the net.*
QNY*	Shift to another frequency (or to _____ kHz) to clear traffic with _____.
QNZ	Zero beat your signal with mine.

*For use only by the Net Control Station.

Notes on Use of QN Signals

These QN signals are special ARRL signals for use in amateur CW nets *only*. They are not for use in casual amateur conversation. Other meanings that may be used in other services do not apply. Do not use QN signals on phone nets. *Say it with words*. QN signals need not be followed by a question mark, even though the meaning may be interrogatory.

Appendix B

The Considerate Operator's Frequency Guide

The following frequencies are generally recognized for certain modes or activities (all frequencies are in MHz) during normal conditions. These are not regulations and occasionally a high level of activity, such as during a period of emergency response, DXpedition or contest, may result in stations operating outside these frequency ranges. Nothing in the rules recognizes a net's, group's or any individual's special privilege to any specific frequency. Section 97.101(b) of the Rules states that "Each station licensee and each control operator must cooperate in selecting transmitting channels and in making the most effective use of the amateur service frequencies. No frequency will be assigned for the exclusive use of any station." No one "owns" a frequency.

It's good practice — and plain old common sense — for any operator, regardless of mode, to check to see if the frequency is in use prior to engaging operation. If you are there first, other operators should make an effort to protect you from interference to the extent possible, given that 100% interference – free operation is an unrealistic expectation in today's congested bands.

Frequencies (MHz)	*Modes/Activities*
1.800 – 2.000	CW
1.800 – 1.810	Digital Modes
1.810	QRP CW calling frequency
1.843 – 2.000	SSB, SSTV and other widebandmodes
1.910	SSB QRP
1.995 – 2.000	Experimental
1.999 – 2.000	Beacons
3.500 – 3.510	CW DX window
3.560	QRP CW calling frequency
3.570 – 3.600	RTTY/Data
3.585 – 3.600	Automatically controlled data stations
3.590	RTTY/Data DX
3.790 – 3.800	DX window
3.845	SSTV
3.885	AM calling frequency
3.985	QRP SSB calling frequency
7.030	QRP CW calling frequency
7.040	RTTY/Data DX
7.070 – 7.125	RTTY/Data
7.100 – 7.105	Automatically controlled data stations

7.171	SSTV
7.173	D – SSTV
7.285	QRP SSB calling frequency
7.290	AM calling frequency
10.130 – 10.140	RTTY/Data
10.140 – 10.150	Automatically controlled data stations
14.060	QRP CW calling frequency
14.070 – 14.095	RTTY/Data
14.095 – 14.0995	Automatically controlled data stations
14.100	IBP/NCDXF beacons
14.1005 – 14.112	Automatically controlled data stations
14.230	SSTV
14.233	D-SSTV
14.236	Digital Voice
14.285	QRP SSB calling frequency
14.286	AM calling frequency
18.100 – 18.105	RTTY/Data
18.105 – 18.110	Automatically controlled data stations
18.110	IBP/NCDXF beacons
18.162.5	Digital Voice
21.060	QRP CW calling frequency
21.070 – 21.110	RTTY/Data
21.090 – 21.100	Automatically controlled data stations
21.150	IBP/NCDXF beacons
21.340	SSTV
21.385	QRP SSB calling frequency
24.920 – 24.925	RTTY/Data
24.925 – 24.930	Automatically controlled data stations
24.930	IBP/NCDXF beacons
28.060	QRP CW calling frequency
28.070 – 28.120	RTTY/Data
28.120 – 28.189	Automatically controlled data stations
28.190 – 28.225	Beacons
28.200	IBP/NCDXF beacons
28.385	QRP SSB calling frequency
28.680	SSTV
29.000 – 29.200 AM	
29.300 – 29.510	Satellite downlinks
29.520 – 29.580	Repeater inputs
29.600	FM simplex
29.620 – 29.680	Repeater outputs

ARRL band plans for frequencies above 28.300 MHz are shown in *The ARRL Repeater Directory* and on **arrl.org**.

Appendix C

The following appendix is reprinted with permission from **parksontheair.com**. For the complete rules, a FAQ, and tons of additional information, please visit the website.

Basic POTA Rules

At a foundational level, POTA is about radio operators visiting parks and making contacts from within the parks with other radio operators at any location. The following sections go into more detail, formally defining this.

Key Definitions

Activator: An activator is a licensed amateur radio operator in a park on POTA's designated list who contacts other licensed amateurs.

Hunter: A hunter is any other licensed radio operator who contacts an activator at a park. The term *chaser* is sometimes synonymously used.

Activity: The act of an activator visiting a park intending to operate POTA is termed an *activity*.

Activations / Attempts

- A successful activation requires a minimum of 10 QSOs from a park in the designated list within a single UTC day (Zulu day).
- Courteous activators will still submit logs for unsuccessful activations to ensure their hunters get credit for the QSOs.
- Multiple activities at the same park in the same state/province/entity and the same UTC day count as a single activation, provided that the 10 or more QSOs combined were made.

Eligible QSOs

1. **Bands/Modes:** All Bands and Modes available to the activator based on their license class may be used in Parks on the Air, according to the details specified in the **logging requirements** section of this document.
    - **Note:** POTA is not a contest; POTA QSOs may take place on any amateur band, including the WARC bands (30m/17m/12m).
2. **Land repeaters are not allowed**: Contacts made via land repeaters are not permitted.
    - Refer to **Parks on the Air's glossary** for the definition of a Land Repeater as it pertains to POTA.
    - You may use a repeater to request direct contacts, but the QSOs submitted for activator credit should not be via a land repeater.

3. **Satellite repeaters are allowed**: Satellite contacts are permitted.

- Refer to **Parks on the Air's glossary** for the definition of a Satellite Repeater as it pertains to POTA.
- Logs should be submitted with the band information for the activator's transmit frequency.

4. **Spotting**: You may self-spot yourself on the **POTA spotting page;** anyone (including hunters) can also spot/respot you, regardless of whether they are working that activator.

- Spots will disappear from the spotting page approximately 30 minutes after the last spot or if the spot comment marks QRT.
- If you are activating a park but also leveraging the QSOs for getting credit for other programs, e.g., operating from a park for ARRL Field day, there may be rules against self-spotting in the other program. Please operate in the spirit of the programs.

5. **Power limits**: POTA does not have a power limit. However, you must still adhere to legal limits based on your license class/band plans and use the minimum transmitter power necessary to carry out the desired communications.

6. **Fully automated QSOs are prohibited**: Each contact must include direct action by both operators making the contact. Fully automated contacts are prohibited.

Activation Location and Access

1. Activations must be performed from parks in **POTA's designated list**, which are also open to the public.

- A park is considered open when the public has civil and legal access, or a special permission/permit to the public lands defined by the park boundary map is obtainable. Such access must comply with any other specific civil or legal restrictions mentioned by the governing agency/website.
- A park is closed when the governing agency/website clearly says that public access is prohibited or when the park itself ceases to exist.
- Seasonal closure of facilities, concessions, offices, some gates, or even large portions of the park may still allow lawful access to the remainder. If in doubt, please call the park office/administrator.

2. The activator and all equipment must be within the park's boundary and on public property.

- Use maps provided by POTA as guidelines only. Refer to official park websites and agencies to find the official boundaries.
- Activations from vehicles, RVs, etc., parked on public property within the park's boundary are permissible.
- Aeronautical activations are permitted if the QSOs are made from airspace directly above the park.

3. Activators may not trespass on private property to access state/provincial, or federal lands.

4. Activators may not attempt to activate from private property, even if the private property is adjacent to, or surrounded by, park property.

5. If a trail system or a river is designated as a POTA entity by itself (not as parts of a land park having a defined boundary), the activator and the station equipment must be on public property within 30.5 meters (100 feet) from the edge of the trail or river.

6. Activations of multiple references ("Multi-loc" or "two-fer," "three-fer," etc.) are permitted with POTA.

- Activation Location and Access Rule #1 applies to simultaneous activations. Such a multipark activation requires an overlapped area where all activated parks' boundaries intersect. The intersection must entirely contain the activator and the station equipment.
- A separate log must be submitted for each park of the multipark simultaneous activation

Hunter Location

1. The hunter can contact the activator from home or any other property or station.
   - If a hunter is also at a **designated POTA park**, this QSO becomes a "Park to Park" (P2P). All other activator rules have to be followed for it to be a valid park to park QSO. See our **Park to Park** page for more info.

Logging Requirements

1. Logs have to be submitted following the logging requirements documented below to count as a valid activation.
   - Hunters do not submit logs to the program. Only activators do. Hunters earn credit through the activator's logs.
2. There are no time limits for log submission.
   - Courteous operators upload their logs without excessive delay, as hunters depend on the activator's log for credit/awards.
3. One log can cover multiple activity days but should be for only one park in a single location(in case the park straddles multiple locations). This also applies to club and multiple-operators activities.
4. Logs must be in ADIF format, with exceptions identified at the end of this section.

Additional notes / guidance

Parks on the Air does not require a specific exchange in a QSO.

Qualification of Parks, Trails, and Rivers for Addition to Parks on the Air

1. All new parks must be owned and operated by a State/Provincial or Federal/National agency. We do not include parks that have part/shared ownership, sponsorship, or are operated by private organizations or local governments
2. To be considered for inclusion, the park must meet the above requirements and offer an informational website detailing current boundary maps.
3. POTA does not create sub-parks within existing parks if both parks are owned and operated by the same State/Provincial or Federal/National agency.

Please be aware that we are temporarily holding the entry of most new parks in the US, with an exception for newly commissioned parks.

Reporting Violations of the Rules

Parks on the Air is a self-regulated, community-monitored program. If you observe a violation of these rules that you feel needs to be followed up on, you may report the violation to **help@parksontheair.com**. You must provide enough details to prove without a doubt that the violation occurred before any action will be taken.

Park Access & Information

If a listed park is permanently closed, public access is prohibited per the Park Access Section of these Rules, the park no longer exists, you are told you are not allowed to operate, or any of the information about the park needs to be updated, please report it to **help@parksontheair.com**.

Guidelines, Interpretation & Intent

Parks on the Air has kept its rules simple because the idea is just to get out and have fun. For some individuals, though, this can be somewhat ambiguous, so this document section provides guidelines on POTA administrators' and developers' intent when setting up the system. Deviating from these guidelines won't get you in any trouble, but it may cause your stats and awards to behave differently than what you and others may expect.

Park Boundaries and Multi-Park Activations

Please be aware of the beauty in the simplicity of the rule "the activator and all equipment must be within the park's boundary and on public property." If an activator is straddling the park or state lines, they are, therefore, not fully within the boundary and are out of compliance with the rule. Similarly, for multipark activations, the activator and all equipment must be entirely within the bounds of each park being claimed, so multipark activations can only occur if the activator is physically located in an intersection where all the claimed parks overlap.

Contributors

Bill Brown's, K4NYM, CW experience goes back to his time in the Air Force as a Morse Intercept Operator. He lives in Central Florida with his wife, Laurie, and Pembroke Welsh Corgi, Mr. Darcy. You can find him on Twitter @POTAactivator.

John Ford, ABØO, was first licensed in Canada in 1981 and wrote the US Tech, General, and Extra in one sitting in 2003. Educated as an Electrical Engineer, most of his ham radio operating has been portable or from restricted HOAs, and he has built almost every antenna he has used over the last 40 years. He enjoys researching antique ham radio books and magazines for portable antenna ideas.

Jherica Goodgame, KI5HTA, is 19 years old and studies English Education and Classics at the University of Mississippi. She was licensed in 2019 but has been active in amateur radio with her father most of her life. She first discovered her love for Parks on the Air in 2020, during the height of the COVID-19 pandemic. She not only hopes to teach but to help integrate amateur radio and wireless technology back into schools after college.

Matt Heere, N3NWV, is an Extra-class operator who was first licensed in 1993. He currently handles public relations and social media for Parks on the Air and is a regular park activator. Matt is an electrical engineer, and his YouTube channel (**youtube.com/MattHeere**) is a collection of build projects ranging from wood working to antennas to "pixel" Christmas lights. He is a private pilot, certified scuba diver, and married with two teenagers at home.

Pete Kobak, KØBAK, was a latecomer to ham radio in his mid-50s. Interested in mobile and portable operations from the start, he initially participated as a mobile op in state QSO parties and VHF contests. NPOTA, followed by POTA and CNPOTA, expanded his mobile operating opportunities. Pete converted a TV production van into a rover for VHF contesting. He is a grateful member of the Mt. Airy VHF RC (The PackRats) and the Pottstown Area Amateur Radio Club.

Harold Kramer, WJ1B, was first licensed in 1962 as KN1ZCK. He was the Publisher of *QST* and Chief Operating Officer of ARRL from 2005 until his retirement in 2016. He began his career in the emergent cable television business in 1973 and remained in the cable television business until 2000 when he became Chief Technology Officer of Connecticut Public Broadcasting. He holds degrees from the University of Connecticut and the University of Hartford.

Sean Kutzko, KX9X, is the former ARRL Contest Branch Manager (2007 – 2013) and Media and Public Relations Manager (2013 – 2017). He was the creator and co-administrator of ARRL's National Parks on the Air (NPOTA) program in 2016. Licensed since he was 14, Sean spent years as an HF contester and DXer before embracing portable operating. You can find Sean on most satellites, running FT8 on the VHF/UHF bands, and activating rare grid squares and parks as he enjoys "the other side of the pileup." He lives in Champaign, Illinois.

Lisa Neuschler, KC1YL, got her tech license in 2015, general and extra in 2016. Her ham adventures started at the Greater Norwalk Area Radio Club (ARC), Norwalk CT. When she retired and moved to Florida, she joined the St. Petersburg ARC and is currently its Vice President. In her free time, she drives around activating parks with the Parks on the Air program. POTA ON!

Clint Sprott, W9AV, was first licensed as KN4BOM in 1955 at the age of 12 in Memphis, TN. After receiving a BS in physics from MIT in 1964, a PhD in physics from the University of Wisconsin in 1969, and 35 years on the physics faculty, he retired in 2008 and rekindled his teenage radio passion. He continues his research in nonlinear dynamics and has authored 14 technical books mostly on chaos and electrical circuits.

Kevin Thomas, W1DED, is an Amateur Extra-class operator who recently returned to ham radio after getting his first license decades ago while a high school student in northern Maine. When he's not selling art via his Portland-based art business, you'll find Kevin perfecting his understanding of operating portably, honing his contesting skills with the help of capable mentors, or learning more about ham radio by interviewing "Founders, Leaders, and Legends" on his YouTube channel (**youtube.com/@w1dedworldwidehamradio**).

Thomas Witherspoon, K4SWL, founder of **QRPer.com**, a website specifically devoted to field radio, has been a ham radio operator since 1997 and makes most of his contacts in the field. His YouTube channel (**youtube.com/@ThomasK4SWL**) publishes real-time POTA field activation videos from the operator's perspective. Thomas was inducted into the QRP ARCI QRP Hall of Fame in 2023.

Bob Witte, KØNR, enjoys a wide variety of amateur radio activities, HF through UHF. For Summits on the Air (SOTA), he achieved Mountain Goat Status (1,000 activator points) using only VHF/UHF frequencies. His book, *VHF, Summits and More*, covers his mountaintop operating adventures. Bob has also contributed to *QST, CQ, CQ VHF, QRP Quarterly*, and his personal blog at **k0nr.com**. He has had a long career in electrical engineering and technology management. He currently works as a consultant on electronic measurements and wireless communications.

Kerri Wright, KB3WAV, started in amateur radio in 2011 but it was a very peripheral hobby until the ARRL National Parks on the Air event in 2016. That event motivated her to get out to the parks with a portable setup, and even to consider learning Morse code. Since then, she's broadened her ham radio experience with digital and CW QSOs and various antenna setups, so she can activate parks in all types of band conditions and crazy locations.

Jeff Zarge, K3JRZ, passed his Technician exam in January 2015 and his General in March of that same year. Jeff has explored many facets of amateur radio from ragchewing on his local repeaters to chasing DX on HF. He's also been a Volunteer Examiner for GLAARG and the ARRL since Autumn 2022 and is currently studying for his Extra. Jeff has been active in POTA since January 2021 as an activator and videos his activations for his YouTube channel at **youtube.com/@K3JRZOnTheAir**.

Advertiser Index

Ham Radio Outlet	**hamradio.com**
Radioddity	**radioddity.com**
MFJ Enterprises	**mfjenterprises.com**
Buddipole	**buddipole.com**
Wolf River Coils	**wolfrivercoils.com**
Comet Antennas	**natcommgroup.com**
Icom	**icomamerica.com/amateur**
Bioenno	**bioennopower.com**
Digirig	**digirig.net**
W5SWL Electronics	**w5swl.com**

US Amateur Radio Bands

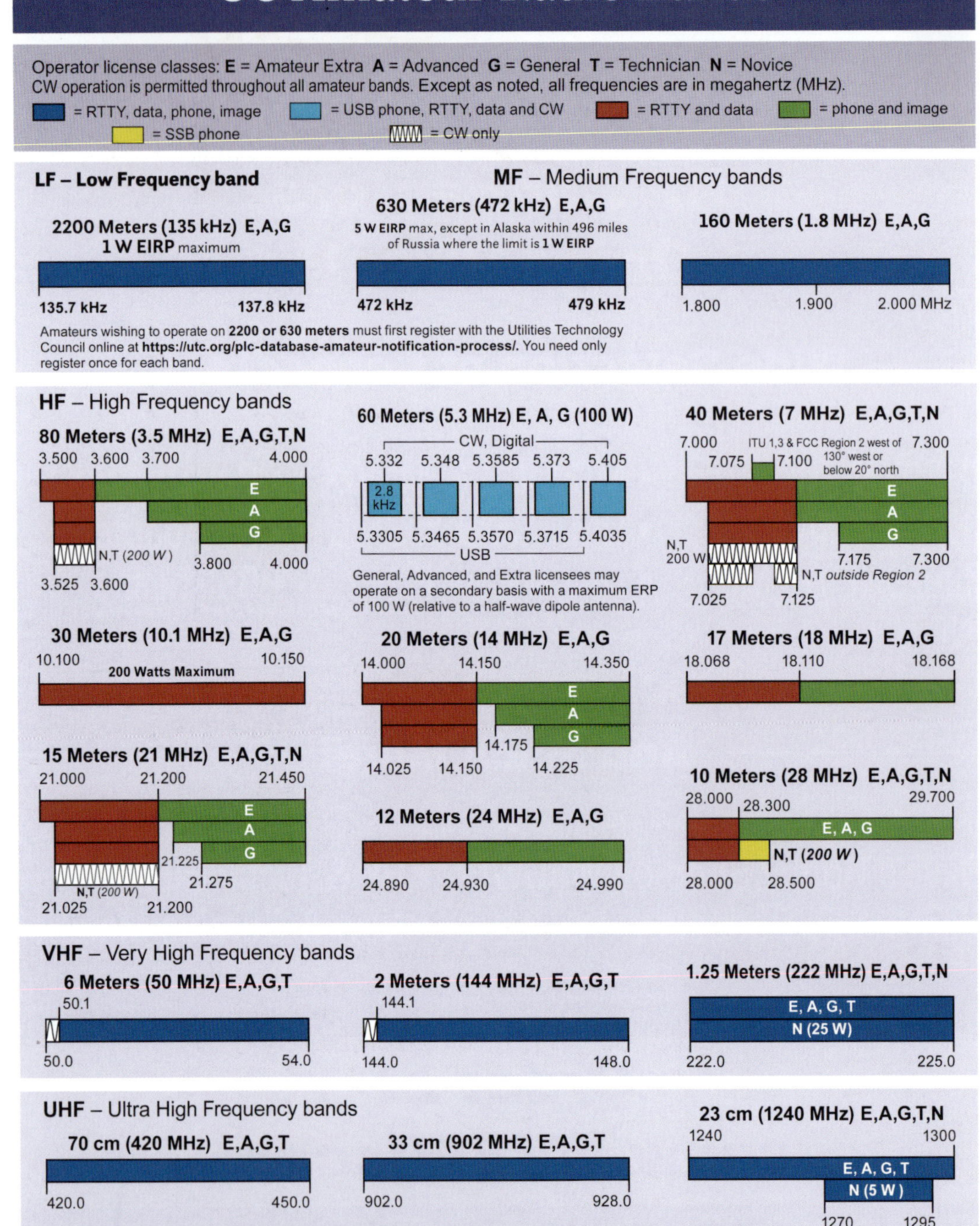

SHF&EHF – Super and Extremely High Frequency bands

All licensees except Novices are authorized all modes on the following frequencies:

2300-2310 MHz	3400-3450 MHz	10.0-10.5 GHz	47.0-47.2 GHz	122.25-123.0 GHz	241-250 GHz
2390-2450 MHz	5650-5925 MHz	24.0-24.25 GHz	76.0-81.0 GHz	134-141 GHz	All above 275 GHz

See www.arrl.org/band-plan for detailed band plans.

OTAbands **rev. 2/10/2023**